THREE BIOLOGICAL MYTHS:

RACE

ANCESTRY

ETHNICITY

Three Biological Myths:

Race

Ancestry

Ethnicity

Alain F. Corcos

Three Biological Myths: Race, Ancestry, Ethnicity

Copyright © 2018 Alain F. Corcos. All rights reserved. No part of this book may be reproduced or retransmitted in any form or by any means without the written permission of the publisher.

Published by Wheatmark®
2030 East Speedway Boulevard, Suite 106
Tucson, Arizona 85719 USA
www.wheatmark.com

ISBN: 978-1-62787-584-4 (paperback)
ISBN: 978-1-62787-585-1 (ebook)
LCCN: 2017961075

Contents

Preface		ix
Introduction		1
1	The Blood Theory and the Jig-Saw Theory: Erroneous Concepts of Ancestry	7
2	The Gene Theory of Heredity: A Better Concept	19
3	The Myth of Human Races	33
4	Ancestry: A Biological Myth	43
5	Ethnicity: A Biological Myth	54
6	Genetics and Jewish Identity	64
7	Each of Us Is Biologically Unique	75
8	Nature and Nurture: You Cannot Have One Without the Other	86
9	Two Revolutions in Biology	99
10	The Biology of Skin Color	106
11	What's the Use of Race or Ethnicity?	128
12	Conclusion	134
Appendix 1: Genetic and Quantitative Aspects of Genealogy		139
Appendix 2: What Is Mitochondrial DNA?		141
Selected Bibliography		143

To my parents, Maurice and Simone Corcos, who contributed half of their genes to me, and to my daughters, Christine and Isabel, to whom I contributed half of my genes!

Preface

When my daughter Isabel was born, her mother told me that she was worried because the number on her little hospital bed was not the same as the one on her wrist. So I, as a geneticist, went to check if she was our daughter, I did not use blood typing or DNA analysis, As soon as I saw her, I knew she was my daughter. She had one of my physical traits. I have prehensile toes. I can pick up my socks with my toes. Isabel was playing with her sheets, the same way as I play with my socks. Reassured, we could go home.

Parents transmit traits to their children. Nothing extraordinary here. But what is not so well known is the part of the grandparents. Because they are four grandparents, they believe they contribute one fourth of the genes to their grandchildren. It is far more complicated than that because of the biological process meiosis which throw a monkey wrench in the concept of ancestry. In the next chapters I am going to explain why there is no link between you and your ancestors six generations back.

Introduction

Every ethnic minority in seeking its own freedom, helped strengthen the fabric of liberty in American life.
                                        John F. Kennedy

I came to the United States for the first time during Word War II as a member of the French Air Force and the second time as a student in 1947. I realized a long time ago that I would never be completely Americanized. There are always things in my behavior that betray my origins, such as my accent, taste in foods, types of entertainment, lack of interest in college sports and school spirit. I also realized that I was not a typical immigrant, I was fortunate to become a university professor, skipping—like many scientists, engineers, and physicians, who came from Europe or Asia—the trauma generally associated with immigration. I became about eighty percent American.

Although I am not completely Americanized, my daughters are. No one can tell that they have a French father, unless they speak French. As to my grandsons, they were perfect American teenagers. Living in San Diego, they did not learn

French in high school, but Spanish.[1] They are Americans, not French-Americans or German-Americans, as their family name Stockhoff suggests.[2]

In other words, it is impossible for the first generation of immigrants to be wholly Americanized. However, it takes no more than three generations for their families to be. For the most part, the United States is the home of descendants of immigrants. Many of the children of immigrants are bilingual, but the grandchildren are not and have little notion of their ancestral culture. They are completely assimilated. Culturally, my American family is the result of the U.S. melting pot, which in spite of its critics, is still boiling. The notion of the melting pot is that the American environment transforms immigration cultures into a specific American one.

Once a French journalist stayed two weeks in New York and had the nerve to write an article on American culture. I have been in the United States for more than seventy years and I can't say I know everything about it. However, I found that the differences between American culture and the culture of other countries are great and the differences between the ethnic cultures within the U.S. are very small, if they exist at all.[3]

---

1 When my grandsons were very young I read them stories in French. I doubt that they understood any of it, but if I skipped a page, they noticed and told me in English not to do it again.
2 Strangely, no one speaks of French-Americans, possibly because few French people have recently immigrated in the U.S. However, before the French Revolution, many came to the territory which became the state of Michigan and gave French names to cities and towns. Pontiac, Au Sable, Detroit, and more. French family names exist among the Michigan population, for example, Delacroix, Devereaux, and Cordier. We can of course find the same names in Louisiana whose capital has a funny name, Baton Rouge—Red Stick.
3 One striking difference has to with the number of protestant

It seems to me that the last bastion of ethnicity is food. There are Italian, French, Chinese, and Middle Eastern restaurants. But often the food they serve is not authentic, but adapted to the American taste.[4] The owners of these restaurants are not always from the nations that the name of the restaurant suggests. Once, I ate in an Italian restaurant in the upper peninsula of Michigan where no one could speak Italian. Another time I ate in a restaurant with a French name in Escondido, California; the owner did not speak French, but the cook did. He had been in the United States for only nine years and was very happy to speak his native language with me. This reminds me a joke of a colleague of Chinese ancestry. We asked him, "What's the difference between a Chinese and a Japanese restaurant?" He answered: "The cook."

It is common to speak about European immigrants as ethnic groups, depending from where they came from. Although they brought different cultures, they could not be physically distinguished. The main physical characteristic that distinguishes some Americans from others has always been skin color, the mark of servitude and object of racism. In this respect, and only in this respect, blacks might still be looked at as a distinct group.

There was always a whittling away of the collective existence of the original "ethnic" groups: the immigrants. The factors involved in this process are marriages between so-called ethnic groups and the fast assimilation of children in the American culture. A high rate of "intermarriage" signals that individuals no longer perceive cultural differences sig-

---

churches in this country. It seems that there is one with a different name on every street corner of a big town.
4   This is also true in France about Chinese Restaurants.

nificant enough to create a barrier to a long-term union. The greatest rate of intermarriage is between people of European ancestry. Its lowest rate is between people of European ancestry and African ancestry. But, intermarriage is on the rise. According to a recent Pew report in 2015, one in six newlyweds had a spouse of a different ethnicity--that is more than five times the number of intermarried couples in 1967. The most common intermarriage pair is "Hispanic" and "White." The next one is "White" and "Asian."[5] The significance of intermarriage goes beyond the loss of some members from the ethnic group. It raises serious concerns about the future of these groups as distinct social entities.

    It is fortunate that parts of immigrant cultures have been introduced with success into the American life; for example, Mardi-Gras in Louisiana. Mardi-Gras is a season of parades and masquerade balls and king cake parties. The carnival has its origin in Southern Italy and Southern France where it is a good occasion for young men and women to meet, throwing flowers at each other.

    In spite of Mardi-Gras, I found that the American culture is a lot different from the French culture and comparing cultures around the world, as one of my retired friends does, is very interesting.

    However, as a biologist, I am puzzled by the fact that some people believe that ethnicity has some biological basis. It does not. Biology destroys the concept of ethnicity because there are no ethnic chromosomes or genes. The concept of ethnicity, like the one of race, is a human invention. Ethnic groups are populations of human beings whose members identify with each other, on the basis of a real or presumed

---

5 Ashley, May. When tit comes to marriage, race and ethnicity matter less. USA Today, May 18,2017

common genealogy or ancestry. That is, ethnicity is a shared cultural heritage: a sense of history, language, and religion. It refers to a decision people make to depict themselves or others as the bearers of a certain cultural identity. Ethnic differences are not biologically inherited; they are learned. Genes help us to acquire a culture, but do not determine which culture we will acquire.

If ethnicity is a social trait that has nothing to do with biology, ancestry seems to have everything to do with it. After all, you are descended from your ancestors. Without them, you would not be alive. This is obvious. But what did you inherit from your ancestors? You will say: genes, chromosomes, DNA. Yes, but are they specific to your family? The answer will surprise you.

Everyone, including the author is interested in his or her ancestry. One of his uncles was able to trace his ancestry to the French Revolution. The father of his son-in-law was able to trace his ancestry as far back as 1641. In recent years it has been possible to trace one's ancestry by using another approach: genetic genealogy that compares your DNA with the one of others, however, these DNA companies promise more than they can deliver.

Twenty years ago, I wrote a book, *The Myth of Human Races*.[6] In it I demonstrated that the idea that human races exist is a socially constructed myth that has no grounding in science. Regardless of skin, hair, or eye color, stature, or physiognomy, we are all of one race. Ten years later I realized that ethnicity was also a biological myth. Although many books

---

6   Alain F. Corcos. *The Myth of Human Races* (East Lansing, MI: Michigan State University Press, 1997

have been written on the biological myth of human races since mine,[7] none seemed to discuss the biological myth of ethnicity. In 1981 Stephen Steinberg wrote *The Ethnic Myth*,[8] but his book speaks directly to current cultural, social, and political difficulties in the construction of a multiethnic society. Nowhere Steinberg is mentioning that ethnicity has no biological basis. The closest he comes to this concept is in the chapter, "The New Darwinism."

Today human diversity is the talk of the town. We want diversity on college campuses, among jury members, police, and everywhere. But very few of us know the biological reason of our diversity. The politicians and the Supreme Court justices do not know it. And yet, Congress passes laws concerning race and ethnicity and the Justices rule if they are constitutional. Worse, the reason for our biological diversity is not taught in biology classes. Would you believe that our diversity (and the one of any organism) is in the formation of sex cells?

---

7 Second edition (Tucson, AZ: Wheatmark, 2016)
   Joseph Graves, The *Race Myth: Why We Pretend that Race exists in America* (New York, Dutton, 2004); *Robert* Wald Sussman. *The Myth of Race* (Cambridge: Harvard University Press, 2014); Michael Yudell *Race Unmasked* (New York: Columbia University Press ,2014); Sheldon Krimsky and Kathleen Sloan (eds) *Race and The Genetic Revolution: Science, Myth, and Culture* (Columbia University Press, 2015); Ian Tatersall and Bob Desalle. Race? *Debunking a Scientific Myth* (Texas University Press, 2011).

8 Stephen Steinberg. *The Ethnic Myth* (Boston: Beacon Press, 1981)

1

The Blood Theory and the Jig-Saw Theory: Erroneous Concepts of Ancestry

The Blood Theory

> Before 1900 it was thought of [the transmission of heredity] as the passage of something from the parents which, like a fluid substance, could mingle and blend in the offspring. The contribution of each parent, popularly referred to a "blood," was assumed to lose its own individuality in the blend which occurred in the child, and this blending process repeated itself in the children's children and in later decedents.
>
> L.C. Dunn

Many people still believe that we are a mixture of the blood of all our ancestors; no matter how far back an ancestor is, some of his/her "blood" flows in our veins. Statements such as this are not only completely false, but have also led to the most violent racism. The fact is that no one else's blood—

only what we produce ourselves—flows in our veins. Blood carries neither traits nor characteristics, such as blue eyes or blond hair.

Yet, for centuries, the blood theory of inheritance has been responsible for the victimizing of at least two ethnic groups: Blacks and Jews. These groups were often killed because of their skin color or religion. Today they are still victims of racism. So, the questions of who is black and who is Jewish continue to be of utmost importance. However, the answers to these terms are far from being clear.

Who Is Black?

> A Jew can change his religion and his name, but a black man cannot change his skin color.
> —A black colleague of the author.

How many black ancestors do you need to be considered black? In the past, it depended on which state you were in. In South Carolina, you were legally black if one of sixteen ancestors were black; in Louisiana, one of thirty-two. But today, if your skin is dark, you are considered black, whatever your ancestry is.

This definition of who is black is based on the one-drop rule, which means that a single drop of "black blood" makes a person black. However, of course, there is no such thing as black blood or white blood. Our blood is red whatever our ancestry. Yet, during World War II, forty years after the birth of genetics, the American Red Cross segregated the blood given by blacks from that given by the rest of the population because it was feared that through blood transfusions, characteristics such as skin color from "Negroes" would be transferred to "non-Negroes." One wonders how many GI's, both black and white, died during World War II because they were prevented from getting the right blood type because of racism.

Although the one-drop rule was shown to be a myth scientifically, it is still firmly entrenched in both the black and white communities and in American custom and law. The blood theory persists in the minds of some people including lawyers[1] and social workers.[2] As late as 1949, dictionaries described Negroes as members of the Negro race, a person having more or less Negro blood.[3]

It is ironic that today some blacks support the one-drop rule. Initially the black community had no choice but to accept it and teach its children how to cope with it, but eventually it came to favor the rule, because it wanted to share centuries of racist experiences that resulted in a common culture. Politically the one-drop rule made sense, if the objective was to keep light colored members from defecting to the white community.

Being black is a social trait that cannot be biologically inherited. We inherit genes for physical traits, such as skin color. Yet, this is often not clear, even to scholars such as F. James Davis. In his book, *Who Is Black? One Nation's Definition* he writes, "The race for which [Walter] White negotiated was the ethnic group with which he identified *certainly not his correct genetic classification* (italics added).[4] There is no such

---

[1] Sometimes lawyers write wills with the expression, "children by blood."
[2] This is seen in the case of the adoption of children. White parents are sometimes prevented from adopting black children, and vice-versa.
[3] "Negro." *The American Everyday Dictionary* (New York Random House, 1949).
[4] Davis, F. James. *Who is Black. One Nation's Definition* (University Park: The Pennsylvania State University Press, 1991) p.7.

thing as a genetic classification of race. What Davis meant was "certainly not the way White *looks*—his skin was light-colored and his eyes were blue."

Black is a new term; in the past people with dark skin were called Negroes. But this term was abandoned socially because it led to a racist slur and was then dropped in favor of the term black. Regardless, both terms refer to skin color. The term African American is also used; however, it presents problems because some North African people such as Berbers or Copts are light-skinned and not considered black and not all people with dark skin are from Africa. If a light-skinned North African man marries a light-skinned American woman, are the children African Americans? No, because the children are light-skinned and are therefore not considered African Americans. Skin color is the most physically striking and socially confusing human characteristic. It is confusing, because in America, being white or black sometimes has little to do with skin color and more to do with ancestry. Many descendants of slaves are light skinned, because they also have European ancestors from whom they have inherited light-skinned genes (see chapter 10). This brought misery to some people like the singer Lena Horne, who suffered all her life for being a light-skinned 'black' person in school and on the stage where she did not fit the desired image of a black entertainer for white audiences. There is a strong attitude in the black community against passing as white, and the apparent infrequency of permanent passing by those who could do so, suggest how strongly self-perpetuating the one-drop rule is. A person who has always been part of a black family and community cannot turn away from them without experiencing extreme stress and major difficulties. He or she will suffer as long as the color line remains. Hopefully we will overcome this type of racism sooner or later with the recognition that skin color has nothing to do with intelligence, behavior, or anything else.

In the last fifty years, we have made progress, decreasing the importance of skin color; yet, it has not been enough, because a few minutes after President Obama made his first State of the Union address, television commentator Chris Matthews declared:

> "It's interesting: he [Obama] is post-racial, by all appearances. I forgot he was black tonight for an hour. You know, he's gone a long way to become a leader of this country, and passed so much history, in just a year or two. I mean, it's something we don't even think about."[5]

Some of us forgot Barack Obama's skin color since he became president of the United States, and many of us do not care. We judge him as any other president. But to others, skin color (or as they see it, race) still matters. Senator Harry Reid admires Obama, but diminished him, when he portrayed him as "a 'light-skinned' African American, 'with no Negro dialect,' unless he wanted to have one."[6]

When the U.S. Bureau of the Census enumerates blacks every ten years, it uses the nation's cultural and legal definition: all persons with any known black ancestry. In other words, Americans are stuck with the defunct blood theory of heredity. Society is 150 years (or more) behind science. It is

---

5   Calderone, Michael. "Matthews: 'I forgot he was black tonight for an hour'." *Politico*. Capitol News Company, 27 Jan. 2010. Web. 10 June 2016. <http://www.politico.com/blogs/michaelcalderone/0110/Matthews_I_forgot_he_was_black_tonight_for_an_hour.html>.
6   http://www.cnn.com/2010/POLITICS/01/09/obama.reid/

Interesting to note that the one drop-rule does not apply to any other group besides American blacks.

Who Is a Jew?

> Thus, for at least two thousand years of recorded history social actions have been implemented based on the belief that certain people had something inherent in them, something in their blood, that made them less than human and consequently deserving of persecution or even death.
>
> Richard M. Lerner[7]

The blood theory of inheritance reached its apogee in Hitler's Germany. To the Nazis, Jews carried genetic and political diseases in their blood. They threatened the purity of the Volkish blood. And it was Jews and their blood about which Theodor Fritsch wrote his "Ten Commandments of Lawful Self-Defense."

> Thou shalt keep thy blood pure. Consider it a crime to soil the noble Aryan breed of thy people by mingling it with the Jewish breed. For thou must know that Jewish blood is everlasting, putting the Jewish stamp on body and soul unto the farthest generations…Thou shalt have no social intercourse with the Jew. Avoid all contact and community with the Jew and keep him away from thyself and thy family, es-

---

7 Lerner, Richard M. *Final Solutions: Biology, Prejudice, and Genocide* (University Park: Penn State Press, 2010). 11.

pecially thy daughters, lest they suffer injury of body and soul.[8]

Yet, the one-drop rule did not always apply to Jews under Hitler. Fractions of ethnicity played a very important role during the Holocaust, as to who was sent to a death camp. If both your parents were Jewish, you were to be eliminated, but if, for example, only one of your grandparents was Jewish, you were not considered a Jew. If you were a Jew, but your spouse was not, you were not going to a death camp, but to a labor camp where you might die of hunger, disease, or under American or British bombs.

The question, "who is a Jew?" has been asked for centuries. Because Judaism is a religion, the answer should be that those who practice it, are Jews. But it is not that simple, because the answer is interpreted differently depending on whom you ask. According to orthodox rabbis, if your mother is Jewish, you are too, but if your father is Jewish and your mother is not, you are not Jewish. If you or your father have been baptized by an orthodox rabbi, then you can then be considered Jewish. However, baptism by a reform rabbi is not kosher—it does not make you a Jew. On the other hand, according to reformed Jewish rabbis, if your father or your mother is a Jew, you are a Jew. Many American Jews belong to the reform movement and are not considered Jews in Israel because the rabbis there are orthodox and, as I said do not accept baptism by a reform rabbi.

For a biologist like me, Jews belong to a religious federation, no matter how loose it is. It is their religion that separates

---

8 Lerner, Richard M. *Final Solutions: Biology, Prejudice, and Genocide* (University Park: Penn State Press, 2010} 35.

them from the rest of the world, not something biological.⁹ This is not as logical as it seems though, since a civil rights activist group recently made the nonsensical statement: "Not all Muslims are religious." Well, if people are not religious, they are not Muslims, Jews, or Catholics. What the man wanted to say is that "not all Muslims are radical extremists."

Although science has discarded the blood theory of inheritance years ago, it has not disappeared; blacks and Jews are still its victims.

The Jig-Saw Theory

> My son is half a Jew. And this is a problem in the South where if you are a Negro or a Jew, You cannot use the pubic swimming pool. I told him to go in the water only half way.
>
> Groucho Marx

Groucho Marx made that joke a long time ago, but that joke was referring to serious things, such as discrimination that affected so-many people, and the erroneous jig-saw theory of ancestry that many people still believe to be true. We often hear, "I am one quarter English," "one-fourth Irish," "one-eighth German," "one-sixteenth Italian," and so on. In other words, a person is believed to be made up of parts of certain ancestral stocks in given amounts.[10]

---

9 Alain Corcos. *The Myth of the Jewish Race* (Bethlehem, N.J.: Lehigh University Press, 2005)
10 The Jig-jaw theory is not as well known as the blood theory, but it is as wrong and also led to racism. The concept was well described by Amram Scheinfeld in his book *Your Heredity and Environment* (New York: J.P. Lipincott Company: 1965).

ERRONEOUS CONCEPTS OF ANCESTRY

THE "BLOOD" THEORY

THE "JIG-SAW" THEORY

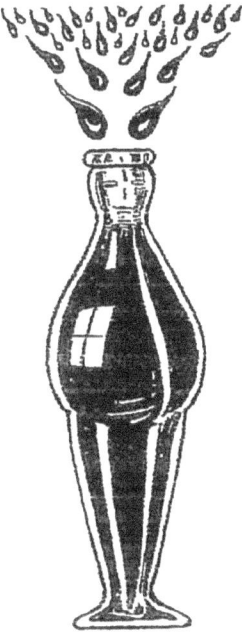
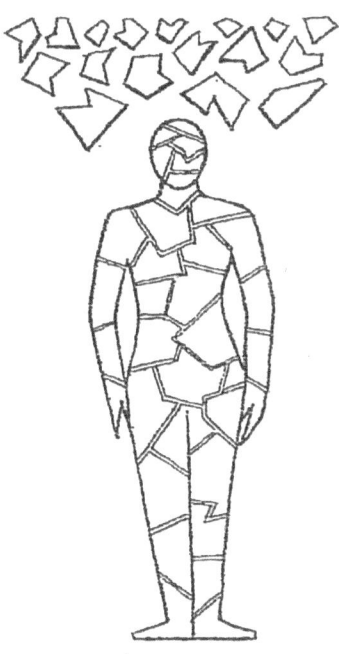

The public still believes in the blood theory. For example, we still use the word "mixed," when thinking of a child of a "black" parent and a "white" parent. The public still also believes in the jig-saw theory. For example, Tiger Woods appeared on the Oprah Winfrey Show, and said that he was not black, but "Cablinasian," because he was the product of Caucasian, African, American Indian, Thai, and Chinese.

However, ancestry cannot be broken down into identifiable ethnic fractions.

These references to someone being precisely this or that fraction of a given ethnic group have no meaning genetically. Nor, beyond one's parents, can anything more than guesses be made as the derivation of one's ancestry in terms of chromosomes, genes, or DNA. As I said previously, the terms English, Irish, German, and Italian refer to nationalities, not anything biological.

In the same vein, one hears "I am half-a Jew" when what is meant is, "My father is Jewish and my mother is not Jewish" or vice a versa. But you cannot be "half a Jew." Judaism is a religion. You either are a Jew, or you are not. Someone asked on the internet:

> "My great-grandfather was a Jew. Does it make me 1/8 Jewish? Last time I checked, I was a Christian. I mean it's not like being white, or being English; it's something you believe in. It has nothing to do with your biology, does it? If one of your parents was Christian, and the other non-religious, would it make you half Christian? I am a 100% Christian. And what my parents, grandparents, and great-grandparents believe in does not make one difference."

Although his knowledge of biology is rough, his idea is correct.

More recently, Elizabeth Dias wrote an article in *TIME Magazine* entitled Evangelicos: "Heber Parades Jr., son of La Roca's pastor preached one Friday evening about the example of the Apostle Paul's assistant Timothy, who was *half*

Greek and half Jew." (Emphasis added).[11] What does this last expression mean? Greek is a nationality and Judaism is a religion. You can be Jewish and Greek, Greek and not a Jew, or vice a versa, but you cannot be half Greek and half Jewish. William Pfaff in an article "Challenge to the Church" in *The New York Review of Books* wrote: "[Gary] Wills has been a critic before of his church, into which he was born in 1934 (half born into it, so to speak, as his mother was Irish Catholic—and a southerner, from Georgia). His father, whose own parents were an agnostic and a Christian Scientist, did eventually become a Catholic, but he was not one while Wills was young."[12] Is Pfaff telling us that Wills is half Catholic—that we inherit religion from our parents like we inherit genes?

If we hear, "I am half a Jew," you do not often hear, "I am half-black or half-white." It is a tradition in this country to classify anyone with darker skin than a European as black. Any child of a white-black couple is considered black. Why not white? What is the basis of such social classification, except racism?

When I told my friends that we cannot be fractions of ethnic groups, some told me: "It is only an expression. We know what we mean." But it is more serious than that. These are expressions of another time, when the concept of race reigned, and when the world was more racist than it is today. It is time to change our terminology. Do not say, "I am half Jewish," when you mean one of your parents is Jewish. Do not say, "I am half-Muslim," when one of your parents is Muslim.

---

11 Elizabeth Dias, Evangelicos, *Time*, April 15, 2013, 25.
12 Http://wwwnybooks com/ articles 20130509/challenge-church/

Today, it is time to abandon both the blood theory and the Jig-saw theory of heredity and accept the gene theory, born more than 110 years ago, which is the basis of genetics, and the real science of heredity.

2

The Gene Theory of Heredity: A Better Concept

Ideas like persons are born, have adventures, and die. But unlike most persons, they do not disappear from this mortal stage; their ghosts walk, often to the confusion of new ideas.
William Wightman[1]

Gregor Mendel's short treatise "Experiments on Plant Hybrids" is one of the triumphs of the human mind. It does not simply announce the discovery of important facts by new methods of observation and experiment. Rather, it is an act of the highest creativity; it presents these facts in a conceptual scheme which gives them general meaning. Mendel's paper is not solely a historical document such as Albert Einstein's on general relativity, but it remains alive as a supreme example of scientific experimentation and profound data.

Mendel's triumph was a lonely one. Neither his fellow members of the Natural History of Brünn, nor the readers of

---

1 Wightman, William P. *The Growth of Scientific Ideas*. New Haven: Yale University Press, 1953. N. pag. Print.

Proceedings,[2] the journal that published Mendel's article, were able to understand at the time (1866) the significance of his achievement. Then, sixteen years after his death in 1884, the development of biology led to the discovery of his findings and interpretations. Three European botanists working independently rediscovered Mendel's paper: Hugo de Vries, Eric Von Tschermark, and Carl Correns. However, only one—Correns—understood the real meaning of the paper in the light of new biological discoveries since Mendel's death.[3] It was common in Mendel's day to regard inheritance as the result of a blending of traits, since it was known that something of this sort occurred in the crossing of certain varieties within plant and animal species. In four o'clock flowers (Mirabilis Jalapa) for example, red-flowered plants produce nothing but more red-flowered plants when crossed among themselves; and white-flowered plants likewise breed true, producing only white flowers. However, when pollen from either a red- or white-flowered plant is transferred to the pistil of the other, the seeds that are formed by this cross produce pink-flowered plants. Thus, a blending of traits could really be seen.

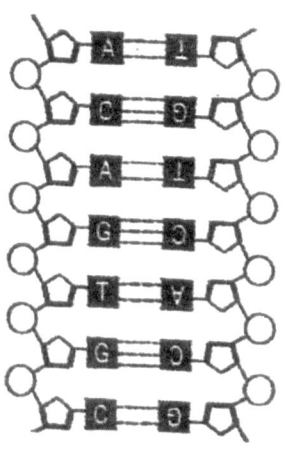

A very small part of a gene. The circles represent the phosphates, the pentagons, deoxyrobose. The interrupted lines linking the bases (squares) indicatehydrogen bonds.

---

2   The Natural History of Brunn was the local biological society where Mendel was an abbot.
3   Alain F. Corcos and Floyd Monaghan. Correns, and independent discoverer of Mendelism. An historical/critical note, *The Journal of Heredity* 78: 330, 1987.

It disturbed Mendel that pink-flowered plants never bred true, as should be the case if a simple blending were responsible. Blending theory of inheritance also failed entirely to explain why hybrids (offspring of parents that differ in a given trait) often revert back to parental types. For example, the offspring of two pink four o'clock flowers may be white, red, or pink. It was equally disturbing to him that this sort of reversion occurred in crosses where no blending was obvious but where a hidden trait kept cropping out.

For his experimental work, Mendel relied upon the garden pea (Pisum sativum), which he knew included several true-breeding varieties that could readily be crossed with each other. Some of these varieties were quite tall and had to be trained as vines, while others were extremely short. Other contrasting characteristics were seed color (green or yellow), seed form (round or wrinkled), and flower position (borne along the main stem or in a group at the top of the stem). In all he worked with seven contrasting traits in this species. As it turned out, no blending of traits occurred. For example, seeds resulting from a cross between tall and short peas did not produce plants that were intermediate in height, but rather, all of them were tall. Mendel called the trait that appeared in the hybrid "dominant" and the trait that did not appear "recessive." Hence tallness proved to be the dominant trait in the cross just cited, and shortness was the recessive trait.

At this point, Mendel took a very important step. He allowed hybrid plants for a given trait to self-pollinate, and he analyzed the results of his seven separate experiments. In each case, individuals showing the recessive trait appeared in a definite numerical ratio to individuals exhibiting the dominant trait. Without exception, in this second generation, a ratio of approximately three dominants to one recessive appeared. For example, Mendel produced 1,064 plants in the tall-short experiments of which 787 were tall and 277 were short. A still further extension of the general experiment revealed that in all seven groups, the individuals showing recessive traits bred true for them, and the remaining, one third of the dominant individuals likewise bred true, and the remaining two-thirds of the dominant did not.

The reason that Mendel was a very successful scientist was because he was not only a botanist, but also a mathematician. He demonstrated this more than once in his paper.[4] To be a mathematician and a biologist is not common, and it is possible that the reason contemporaries of Mendel did not understand his paper and lectures was because they did not understand the mathematical basis of his work.

Mendel then postulated the existence of "characters" that were associated with the gametes (sex-cells) of parent individuals. By allowing letters of the alphabet to represent these characters, he set up his crosses on paper and manipulated them theoretically. For the sake of convenience, capital letters were made to represent dominant characters, and lower case letters were used to represent recessives—a system that is still employed in genetics and plant breeding. Mendel came to the conclusion that only if the genetic constitution of a parent plant were represented by two characters for a given trait, with one and only one of these being transmitted to a gamete (sex cell) could his results be explained.

Mendel discussed characteristics as the unit factors that are transmitted from one generation to the other. It was the Danish biologist Wilhelm Johansen who coined the term "gene" for them. He is also responsible for designating the

---

[4] Corcos, Alain F., and Floyd V. Monaghan, *Gregor Mendel's Experiments on Plant Hybrids*. New Brunswick: Rutgers University Press, 1993. Print.

characteristic of an organism as its phenotype, and factorial basis for the characteristic as its genotype.

Mendel's concept of the gene (although he did not use this term) preceded the discovery of chromosomes. The American cytologist, W.S. Sutton pointed out the similarities in gamete (sex cell) formation between Mendel's characteristics and chromosomes. Soon it was accepted that genes were part of the chromosomes.[5]

Another fruitful branch of genetics—biochemical genetics—came into being in 1937, when George Beadle, an American geneticist, joined forces with Boris Ephrussi, an Italian geneticist, in the study of eye color pigmentation in the fruit fly. The result of their work led them to conclude that each gene is responsible for one primary action during a particular step in pigment formation. Stated more simply, in chemical terms it appeared that a given gene did one thing: produce an enzyme. Later, it was found that a gene is responsible for the production of a protein or hormone.

Although the work of Beadle and Ephrussi led to a crucial genetic discovery, it was handicapped because very little was known about the biochemical pathways involved in the traits they were studying in the fruit fly. This work had paid rich dividends in new insights, but it became apparent that further success in identifying gene action with specific biochemical reactions in this organism would come only with great difficulty.

At this point, George Beadle joined E.L. Tatum, an English biochemist, and they decided to reverse the former strat-

---

5 Mendel was lucky in his choices of traits in his peas. Each one was due to a gene in a different chromosome. If they had not been, Mendel would never have found the mathematical ratios for which he is so famous.

egy. Instead of working from known genes to unknown reactions, they would attempt to go from known reactions to unknown genes and use a simpler organism than the fruit fly. They chose to work with the bread mold Neurospora, since it seemed to be well-suited for genetic studies, and certain biochemical pathways had already been worked out. They also took advantage of the fact that X-rays were now known to induce mutations in genes. They were successful, and on the basis of all their studies, they advanced a conceptual scheme in 1941, that they termed the *one gene, one enzyme hypothesis*. Although this hypothesis was modified later to account for additional data, it constitutes a landmark in the development of genetic concept. Quite deservedly, Beadle and Tatum were awarded the Nobel Prize for Physiology and Medicine in 1958 for this highly significant work they had accomplished around 1940.

By then, the gene was accepted as the unit of heredity and part of a chromosome. But what is the gene made of? The answers came from very ingenious biochemical experiments, which I will not attempt to describe. Suffice it to say that even before 1900 it was suspected that what we presently call *nucleoprotein* (which is composed of protein molecules attached to nucleic acids) was somewhat associated with inheritance. By 1940 the evidence was very strong that deoxy-ribose nucleic acid (DNA) was the functional chemical substance of the gene. DNA is a molecule that does two things: first it acts as the hereditary material, which is passed down from generation to generation and second, it directs, to a considerable extent, the construction of our bodies, telling our cells what kinds of molecules to make and guiding our development from a single-celled zygote to a fully-formed adult. Each human being receives his or her heredity from the union of two bits of living matter—an egg from the mother that has been fertilized by a single sperm of the father. The egg, a tiny sphere, is hardly visible to the naked eye, and a sperm is only visible with the aid of a microscope.

Although male and female sex cells are extremely different in size and shape, an egg is spherical while a sperm is shaped like a thread inflated at one end to form a head, both parents generally contribute equally to their children. How is that possible? In the nineteenth century, some biologists saw that the only reasonable solution of this paradox is to suppose that it is not the sex cell as a whole but some particular bit of it, alike in both sex cells, which is chiefly concerned with heredity. Indeed, a careful study of the sex cell under strong microscopes showed that one part is similar in both the egg and sperm. These similar parts are the nuclei of the sex cells, or more precisely, the chromosomes in the nuclei. Each human egg and sperm contains 23 chromosomes, and heredity is transmitted through the chromosomes.[6] The fertilized egg proceeds to divide into two, four, eight, and eventually billions of cells, which comprise the adult body. Each cell receives, through this division, a nucleus identical to that of the fertilized egg. The heredity received from the parents is, thus, present in every cell of the child.

Genes have different functions. For example, one gene is responsible for the production of the growth hormone. Another is responsible for the production of the beta chain of hemoglobin, the protein that carries oxygen to all of the cells in our body. Although genes are exceptionally stable in composition and function, they can and do change through

---

6   Some maternal heredity is transmitted through mitochondria, the energy organelles, which are transferred from the mother to all her children, boys and girls. We do not inherit mitochondria from our father.

a process called mutation. Genes reproduce themselves and it is during the process of self-replication that mutations may occur. If a gene happens to mutate to a new form, subsequent generations of that gene reproduce themselves in the mutant, or altered form. Such alternative forms of a gene are known as *alleles*. This is now well known, but what is not so well known, is that human diversity is not due to different genes—as many people believe or speak loosely about—but to the different forms of a gene, called alleles. Let us take the example of the inheritance of the ability to taste an artificial substance called Phenyl-thiocarbamide (PTC). One day in 1932, a scientist complained that the powder his colleague was working with had a very bitter taste. The PTC, did not taste bitter to the man actually handling it. Quite by accident, these two had stumbled on what it is now recognized: some people are born with the inability to taste PTC. Their inability to taste PTC is inherited as a recessive gene (t) from both parents while the capacity to taste the bitterness in the compound is transmitted and inherited as a dominant gene (T), T and t are the alleles of the same gene, but (t) is not producing whatever is needed to taste PTC.

Another example is the following: many Africans and about 20 percent of Europeans are born lacking a certain intestinal enzyme, called lactase, necessary for the digestion of milk.[7] This may not become obvious until late childhood or even until adulthood. These individuals suffer diarrhea after drinking milk, but can live normal lives if they avoid milk or drink milk without lactose. Their inability to digest milk is due to a mutation of the L gene; their genotype is ll and those people who are able to digest milk have the genotype

---

7   Daniel Fairbanks *Everyone is African* (Amherst, N.Y. Prometheus Books, 2015).

LL or Ll. Another example is the wax in our ears. Some of us have wet wax while others have dry wax. The different types of wax are due to different alleles of the gene responsible for producing that substance.

But what is a gene? After all, genes had been postulated by the early geneticists and they seemed to behave as if they were always on chromosomes, but which part of the chromosome? Chromosomes are made up mostly of proteins and nucleic acids. For many years, scientists believed that genes were proteins. However, in the 1940's, experiments indicated that nucleic acids, of which there are two types, DNA and RNA, were likely to be the molecules involved in heredity. In fact, in 1953, it was postulated that DNA, specifically, was the chemical of heredity. Its structure was discovered and shown to be able to replicate—it could produce exact copies of itself. This replication is absolutely necessary because when a cell undergoes division, it produces two identical cells. This assumption has led scientists to gain more knowledge about what is going on in a cell, but they still do not know for sure what the exact path is from a gene to a trait.

Genes are located along the chromosomes and each gene occupies a certain place on a chromosome. Since chromosomes occur in pairs, each gene has its counterpart at the same position (locus) on the corresponding (or homologous) chromosome. These two genes can be of identical or different forms. This fact is the first step towards understanding our own diversity, because each of our cells carries ten thousand genes, each of which might be different forms.

Then in 1953, the DNA structure was discovered. It is the now famous double helix of Watson and Crick, a very long twisted ladder, the side rails of which are made up of alternating sugar (deoxyribose) and phosphoric acid groups. The rungs of the ladder are made up of pairs of bases. There are only four bases of DNA—adenine, thymine, guanine, and cytosine—and their spatial structure is such that if one half-rung on the ladder is adenine (A), the other half-rung must be thymine (T) or else the rung will not fit between the side rails. In a similar manner, cytosine (C) must always pair with guanine (G).

AGCATCCGGGGCTTCCTGAGC
Part of the DNA Helix
TCGTACGGCCCCCGAAGGACG

The final element (and a vital one) in the structure of DNA is that the chemical bonds holding the two half-helices—the bonds between G and C and between A and T in the middle of the rungs—are hydrogen bonds. These bonds are very weak, and they allow the two halves of the helix to come apart easily, like the halves of a zipper.

The complimentary structure of two half-molecules is the feature that makes the orderly self-replication of the genes possible (absolutely necessary in cell reproduction). At replication, the two half-molecules come apart and new halves are built up on each of them. Because of spatial restrictions, the new half will always be the perfect complement of the old half. Consequently, two double helices are built up that are exact copies of the original. One has an old right-hand side and a new left-hand side, the other an old left-hand side and a new right-hand side.

old strand AGCATCCGGGGCTTCCTGAGC
new strand **TCGTAGCCCCGAAGGTACG**

new strand **AGCATCCGGGGCTTCCGAGC**
old strand TCGTACGGCCCCCGAAGTGAGC

The linear arrangement of the base pairs along the molecule allows for gene variety. Suppose that a gene in a DNA molecule is 500-base-pair long. Since there are four possible bases at each position, there are 4^{500} different possible arrangements for each; vastly more than the number of atoms in our milky way, not to mention the different kinds of genes we require, which has been estimated at being 30,000.

"The DNA is a code, that is, the order of the bases [the letters: A, T, G, C] makes one gene different from another gene (just as one page of print is different from another)."[8] In other words, the precise sequence of the bases is very important. Any change might bring about a mutation. For example, a base pair A-T changed to a pair G-C lightens one of the genes for skin color (see chapter 10).

The nature of the gene and its function were well known in the 1980s. Genes were part of the DNA, but there was a problem: the DNA was very long and parts of the DNA did not seem to have any function. Geneticists called this part of DNA "junk DNA." This was a mistake because they later discovered that there are genes that regulate other genes or have other functions. The science of genetics has become more complicated because the nature of life is complicated.

I like to end this chapter with the fantastic story of how the genetic code was cracked. I have included this story because it was not only an era of beautiful experiments created and carried out by fantastic minds, but it also led to new avenues in the fields of agriculture and medicine. The genetic code is universal: it applies to every species on earth. It was for the author, a geneticist, an exciting time.

---

8   This is part of a letter of Francis Crick to his son two weeks after the Watson-Crick DNA discovery, but before the publishing of the discovery in Nature (The English Scientific magazine).

How Biologists Cracked the Genetic Code[9]

> The discovery of the [DNA] structure by Crick and Watson, with all its biological implications, has been one of the major scientific events of this century.
>
> Sir Laurence Bragg

I said that genes encode the thousands of proteins that make up our body, but what are proteins? Proteins are long molecular chains made of individual links called amino-acids. Organisms use only twenty different kinds of amino-acids to build their proteins. The blood protein, betaglobin, is a sequence of 146 amino acids in a particular order. If you were to switch any one amino acid, things might go terribly wrong. For example, the normal and sickle cell forms of beta-globin differ by only one single amino-acid.

The next question was, how do the sequencing of A, T, G, and C in DNA determine the sequence of amino acids in proteins? This code, whatever it is, is the code of life. It is the precise physical connection between what we inherit from our parents and how our bodies are built. Although the coding problem could be stated simply, it could not be solved easily. It was clear that the answer could not be simple because the four letters of in the DNA language taken singly could only decode four amino-acids. For months mathematicians and physicists attempted to solve the problem, but to no avail.

---

9 See Matthew Cobb, *Life's Greatest Secret: The Race to Crack the Genetic Code* (New York: Basic Books, 2015) Cobb is a professor of zoology at the University of Manchester, England, He is not only a working geneticist, but also an historian of biology.

Fortunately, speculation about the code did not occur in an empirical vacuum. Biologists discovered that DNA was not the physical template on which proteins were built. DNA did not even interact directly with the protein that it encoded. Instead, an intermediate molecule seemed to be involved. This intermediate molecule proved to be RNA, a close cousin of DNA. RNA, ribose nucleic acid, is single-stranded, and does not have the base T, but the base U. Biologists realized that the sequence of letters in one strand of a DNA encoded for a matching sequence on a RNA molecule. If a DNA sequence reads ATAGCAG, the matching RNA sequence reads UAUCGUC. It was this RNA sequence that then (somehow) determined the sequence of amino acids in a protein.

Despite these developments, the code itself remained undeciphered. Which letters of DNA code accounted for which amino acid? Biologists were no nearer to the answer than they were immediately after the discovery of the double helix. The coding problem had entered a confused phase. But the confusion did not last long. The code was brilliantly cracked by a team unknown at the time: Marshall Niremberg and Heinrich Matthaei of the National Institute of Health in Bethesda, Maryland. They used artificial RNA sequences that carried the same letter over and over—UUUUUU.... and asked what kind of protein was made (this was done in a test tube). The answer, in the case of UUUUUU..., was a protein that carried the amino acid phenylalanine and nothing else. Further experiments carried by sharp minds allowed the deciphering of the entire genetic code. The code was a "Triplet." Every three letters of DNA specify one amino acid. Given a sequence of letters, biologists were then capable of knowing which amino acid would appear and what protein would result. Niremberg was awarded the Nobel Prize in 1968. He was not the only one during this era. It seems that every year in the 1960s, 1970s, and 1980s a molecular biologist was a Nobel Prize winner for medicine.

That period brought more discoveries. One of them, perhaps the most fundamental from an evolutionary point of view was the finding that the code is nearly universal across life on earth (some minor variants of the code exist). All of us—bacteria, fungi, plants, animals, and humans—share the same genetic code because of an ancestor we all share that lived billions of years ago and employed this code.

3

The Myth of Human Races

> *"Race is a term for a problem, which is created by social factors and social factors only, it is therefore entirely a social problem.*
>
> Ashley Montagu

Few concepts are as emotionally charged as that of race. The word conjures up a mixture of associations—culture, heredity, subjugation, exclusion, and persecution. Such racism is based on the idea that human races exist and that some are superior to others. For more than three centuries scientists (or rather pseudo-scientists) have tried in vain to prove the trueness of these ideas.

They have based their flawed research one or more specious assumptions; (1) Humanity can be classified into distinct groups using identifiable physical characteristics; (2) these characteristics are transmitted through "the blood" (3) Characteristics, such as skin color and curly hair, are inherited together (4) physical features can be linked to behavior; (5) human groups, or races can be ranked in order of intellectual, moral, and cultural superiority.

Today, anthropologists and geneticists have demonstrated that there are no human races. As early as 1922, Theophile Simar, in *La doctrine des races*, expressed the opinion that the concept of race lacked all scientific validity and was devised for political purposes.[1] Magnus Hirshfeld, while fleeing the Nazis, died in France in 1935, leaving a book unpublished. His manuscript, *Rassismus* was translated into English under the title *Racism*. In it we find these words:

> If it were practicable, we should certainly do well to eradicate the term "Race" as far as subdivisions of the human species is concerned.[2]

Similarly, cultural historian Jacques Barzun wrote these words in 1937:

> The particular end or object of this work is to show how ill-founded are the common place and the learned views of race.[3]

Although Hirschfeld's posthumous work attracted relatively little attention, Barzun's book was important because it stressed that racism was more universal than previously thought. Racism was not unique to German attitudes toward the Jews, and it could be found in the widespread assumption that the "whites" are unquestionably superior to the "Colored races." In 1940, the prominent cultural anthropologist, Ruth Benedict, wrote *Race, Science and Politics*, which was republished several times thereafter. Her main

---
1 George M. Frederickson Racism: A short history (Princeton University Press, 2002), 159
2 Magnus Hirschfeld. *Racism*. (Fort Washington, 1938)
3 Jacques Barzun. *Race: A Study in Modern Superstition* (London: Methuen, 1938)

concern was to refute the scientific pretensions of believers in racial inequality.[4]

However, it was Ashley Montagu who became the most prominent anthropologist to deny the existence of human races.[5] He considered race to be an utterly erroneous and meaningless concept in the light of genetics, as well as socially harmful. The term "race," he argues, should be dropped from the anthropologist's vocabulary, as "phlogiston" was abandoned by chemists. He played an important role in writing the UNESCO statements on race and racial prejudices.[6]

The term "race" has an interesting beginning. In Spain, the word "raza" was used to mean breeds of horses and dogs. It was also pejoratively used to refer to Jewish and Moorish ancestors. In French, the word "race" seems to have been derived from the Italian word "razza," and has been used since the fifteenth century to mean family, descent or ascent, genealogy, generation. But, in general the word was used only by aristocrats to claim that their heredity of physical and mental abilities was better than the one of everybody else.

The meaning of the word race changed at the dawn of modern times. The concept of human races seems to have had its origin in the events following European voyages of discovery to the New World and beyond. Popular ideas about the differences between peoples grew out of European contact with Native Americans, Africans, and peoples through-

---

4   Benedict Ruth. *Race, Science, and Politics* (New York: Modern Age Books, 1940)
5   *Race and Other Misadventures: Essay in Honor of Ashley Montagu in his Nineteenth Year.* Larry Reynolds and Leonard Lieberman, Eds (New York: General Hall. Dix Hills, 1996)
6   UNESCO, 1967. Declaration on Race and Racial Prejudice stressed the importance of the economic and social causes of racism.

out Asia and the Pacific. Europeans were struck by the fact that the people they encountered appeared to be physically different from them. They could have accepted these differences and treated non-Europeans accordingly. But they did not. They saw them as inferior—people who at least needed to be civilized or at worst, exterminated.[7]

Europeans quickly began to devise systems of classification, but centuries later found that race classification was an impossible task.[8] At first glance, the idea that each human population is characterized by a unique combination of specific traits seems to make sense. It is not hard to distinguish dark-skinned Africans from Australian aborigines or dark-skinned Indians, because each group tends to have some traits in common that others do not have. For instance, Europeans tend to have light skin, straight, or wavy hair, and noses of narrow to medium width. Sub-Saharan Africans tend to have dark brown or black skin, wiry hair, and so on. Another grouping of traits occur among East Asians, most of whom have pale-brown or slightly yellowish skin, straight black hair, and dark brown eyes. A critical analysis of human diversity around the world, however, reveals that this general "racial" view is, at best, simplistic, but is more likely wrong for the following reasons.

First, although we can perceive several main groupings, there are millions of people who cannot be pigeonholed into

---

7 It has been suggested by Thomas Powell that this idea was already present in the minds of Norsemen who came into contact with American Indians five hundred years ago before Columbus. In *the Vineland Stages* the Norsemen named the American Indians Skatelings or wretches. Their first idea of appropriate behavior was to kill them.
8 Alain F. Corcos. *The Myth of Human Races.* (East Lansing, MI: Michigan State University Press, 1997), Chapter

them because they have a characteristic of one group and a characteristic of another group. Second, there is extreme diversity among individuals of the same group. For example, not all Africans have dark skin and not all Asians have epicanthic folds over their eyes. Third, no "racial" trait is restricted to one specific human group.[9] In the case of skin color, many individuals in the world have dark skin. Some live in Africa, others in Australia, and still others in India.

Only in a very loose way are the traits used to classify mankind into races associated with one another. This is indicated by the fact that, when anthropologists draw maps of the incidence of a particular trait around the world, the distribution of a single trait is independent from any other. In other words, if you try to categorize human beings by one trait, such as skin color, certain broad divisions seem to emerge. If you add another trait, such as hair type, the divisions become blurred. The picture becomes even more confused with each additional trait that is mapped. Simple traits such as skin color, hair type, and lip shape are largely independent. They do not cluster to form a particular "racial" type.

People who believe in the existence of human races assume that humanity can be classified into groups according to identifiable physical characteristics. However, no one has been able to find a specific set of characteristics that can be used to distinguish one group from another without introducing layers and layers of ambiguity. Historically, the number of races has varied from four to more than forty. No matter how many groups we propose, we always find people who do not fit any category.

---

9   R. Livingstone (1962) recognized a tautology: traits or genes used to define a population as a race became identified as racial markers.

The reason for this is that race formation never occurred among humans. Race formation requires reproductive isolation and we humans have never been completely isolated. To understand this concept, let us compare humans and our friends, the dogs.

In order to have races of animals, we need to prevent the members of one group from breeding with the members of another group. This is what we did to get the numerous forms of dogs we have today. How did we shape the wolf-like ancestor of the dog into descendants having so many different forms and qualities: a hunting dog such as the English setter, a sheep or cattle driving dog such as the Collie, a guard dog such as the Great Dane, a police dog such as a German Shepherd, or an ornamental dog that does absolutely nothing, such as the Yorkshire Terrier? As the great detective, Sherlock Holmes, would say, "Elementary, my Dear Watson." We did this by artificial selection and complete reproductive isolation.

At first, when we were strictly hunters, we kept all the dogs most useful in the hunt. This was a primitive type of selective breeding. When we domesticated sheep and cattle we needed dogs that could fight off predators, run fast, and were intelligent and obedient. How did we get them? Obviously, no dog had all these qualities. But within the population of dogs at the time, there were some that were not obedient, but ran fast; others were obedient, but slow; others were very intelligent, but not necessarily fast or obedient. Our ancestors thought about a way to produce dogs with all these three qualities. How about mating Rover who runs fast with Rosie who is very obedient? Hopefully, some of their puppies might be both obedient and fast runners, though they might not run as fast as their father, and might not be as obedient as their mother. Nevertheless on the whole, these puppies would be very useful. These dogs would be even more useful if they were intelligent. Mating these dogs with Ginger, who is remarkably smart, would give puppies that could be fast, obedient, and smart.

As long as races of dogs are reproductively separated, we will be able to distinguish them. Every spaniel will look different from any golden retriever. Any Labrador will look different from a foxhound.. This reproductive isolation enforced by us is absolutely necessary. Without it, races of dogs would disappear in a hurry. Think about your favorite highly bred female dog for which you have opened the front door, letting it outside to answer a call of nature! She may encounter a dog of a different race. Sixty-three days later you may be presented with a litter of puppies that share some of the characteristics of both parents. Jack London was right when he wrote the Call of the Wild. The descendants of Buck are running wild as wolves.

We human beings have never been bred and selected by a super authority to produce special races of people. [10] We have never been reproductively isolated. Within bounds, we have been free to choose our mates. In some cases, some of us have traveled thousands of miles to find a spouse, often thanks to the generosity of our governments when engaged in military action. Wars have always offered wide opportunities for breeding between the invaders and the invaded. Even in times of peace attraction between the sexes has often been enhanced by the difference of culture, customs, and manners of the two individuals involved. Sometimes it even overcomes

---

10   The only exception we can think of is during World War II, when Hitler tried to improve the "Aryan race" by selecting blond haired and blue eyed to mate. Hitler, never saw the results of this human selection.

the lack of common language. In this global world, the possibility of race formation in humans will decrease even more since we can be any place on the earth in a few hours.

The idea that there are no human races took a long time to develop. As a matter of fact, it took anthropologists hundreds of years to abandon their efforts to classify humanity. They have finally realized that they were not more successful using blood groups, or other genetic markers, than they had been in the past when they were using skull sizes or skin color. They have finally realized that categorizing human beings into races requires such a distortion of the facts that its usefulness as a scientific tool disappears. Now, they understand their task differently: to study human variability without the concept of race. This idea has been well-defined by Stephen Jay Gould, who answered as early as 1974 to those who declare that race is self-evident:

> Geographic variability, not race, is self-evident. No one can deny that *Homo sapiens* is a strongly differentiated species; few will quarrel with the observation that differences in skin color are the most striking outward sign of this variability. But the fact of variability does not require the designation of races. There is better ways of studying human differences.[11]

In fact, the more scientists dig in the matter, the less they found differences among us and even between us and animals. New genetic discoveries support that idea.

Human populations are not static. They are dynamic, continuously varying in diversity and this is why the word

---

11 Stephen Gould. The Race Problem, *Natural History.* March 1974. 8

"race" is completely inappropriate to use for human populations.

It is American culture that is principally responsible for the perpetuation of the concept of race well after its loss of scientific respectability by the end of the twentieth century. Even the most well-meaning persons, who attempt to grapple with the persistence of inequality between "blacks" and "whites," take it for granted at the outset, that racial categories adequately capture the relevant differences under investigation.

Race should be placed on the scrap heap of outdated scientific ideas. It is time for the rest of us to abandon this obsolete, destructive and false biological notion of human classification. But let us not fool ourselves. It will take time. In his inaugural address at the College of France in 1972, Jacques Ruffie strongly affirmed that, "In man, races do not exist. However, if the idea of race is dead, its burial might take centuries." He was right. Forty years later lawyers, including justices, politicians, media people, still talk about race as if it were a biological entity. Skin color is a biological trait. Race is not.

Many scientists and scholars in the genomic age had joined a consensus that found race to be nothing more than an arbitrary social construction. They had hoped that policies would be enacted to address the existing social injustices that had been created by past false beliefs about race and racial hierarchies. Once this remediation work was done, race would be able to disappear, taking its place among history's misguided and unnecessary concepts.[12]

---

12   When Bill Clinton, Tony Blair, Craig Venter, and Francis Collins, shared a stage in June 26, 2000 to celebrate the completion of the Human Genome Project, they focused on the finding of human

Unfortunately, such social reforms never came to fruition: race continues to be a powerful social force in America, shaping the health, wealth, and political power.[13] Race continues to be a factor in health diagnostics, medical research and in court cases (see chapter 11).

Race still plays a role in commercial fields. DNA tests are sold claiming they are able to recover the ancestral histories of black people whose heritage had been erased by the slave trade. Nike has marketed race-specific jogging shoes designed specifically for the supposedly thicker feet of American Indians, to combat the epidemic of obesity and diabetes on impoverished reservations.[14]

It is going to take decades to understand that blacks and whites are only different because of the amount of melanin in their skin (chapter 10). If political speeches during the presidential campaigns in 2016 are any indication, we are going to wait at least another fifty years, or sadly 100 years to accept this scientific truth.

---

sameness: with all humans sharing 99.9 percent of their genetic code, theories of racial differences or hierarchy could not possibly have any basis in genetics.
13   Joe R. Fagin, *Racist America*. (New York: Routledge:2001)
14   Ian Whitemarsh and David Jones. Governance and the Use of Race in *What's the Use of Race?* Cambridge, MA, MIT Press, p.1

4

Ancestry: A Biological Myth

> *What relatives…share. Ancestor. There is DNA and there are probabilities of sharing some, but no tangible genetic stuff divisible among kin and distinguishing them or bounding them from non-kin. There is no genetic test for kinship. Kinship is not a genetic property.*
>
> Jonathan Marks

What Marks says goes against one of the assumptions that many people make when they are looking for their roots, however, from a biological point of view he is perfectly right. As a matter of fact, there is generally very little biological similarity between our ancestors and us. Some sociologists believe that ancestry is a socially constructed concept as race is.[1] I agree and will explain in this chapter.

Years ago, I was invited to a genealogic society meeting, not because I was knowledgeable in the field, but because I

---

1   Joan H. Fujimura et al. Race and Ancestry: Operational zing Population in Human Genetic Variation Sudies. In *What's the Use of Race?* Ian Whitmarsh and David S. Jones, eds. (Cambridge MA: The MIT Press, 2010), 169

wrote a biographical book,[2] I might give members of the society tips on how to write about their family roots. That particular night, a lady told us that she had finally found the proof that she was a descendant of a passenger on the Mayflower. Not wanting to curb her excitement, I did not tell her that the chances of her ancestor, twenty generations before, transmitting her anything genetic was practically zilch (see table 1).

Yet, we are curious about our family history. We assume that we will understand ourselves better if we know our ancestors and reflect properties within ourselves that were passed down to us by unbroken lines from past generations. We are happy tracing our ancestry a few centuries. I did this on my paternal side, and found that one of my ancestors, Moise Corcos, lived in the city of Bordeaux, France, at the time of the French Revolution. He is my ancestor all right, but the chances that he gave me some of his genes are very small since we are about seven generations apart. I am guessing that one of his ancestors (with the name Corcos) came from Spain from where he and his family were kicked out in 1492, because they were Jews or Conversos. I base my guess on the fact that there is a little town called Corcos in Northern Spain and on the fact that Jews often have family names of towns they came from. However, I like to emphasize that between these Corcos' and I there is no genetic link at all.

Years ago, I was invited to a symposium on Asian-American identity and I had the chance to explain why there are very few links between you and your ancestors. After the speakers had finished their presentations, they asked if there were any questions. A man stood up and asked if he could claim to be a Native American because his great-great-great

---

2   Alain F. Corcos. *The Little Yellow Train: Escape from Nazi France*. Second Edition (Tucson, AZ: Wheat mark. 2005)

grandfather was one. I was invited to answer him and I told him that the chances of inheriting Native American genes, *if these things existed*, were very slim. The notion that all of our ancestors contribute their due share to our genetic make-up is wrong. Did he ever think about the fact that he has a tremendous number of ancestors? He has two parents, four grandparents, eight great-grandparents, sixteen great-great grandparents, and so on. Six generations back, he has sixty-four ancestors. This number is greater than the number of chromosomes (forty-six) he has in his cells. Hence, it is not possible that he received all of the chromosomes from his sixty-four ancestors. Obviously, some had been lost.

The man and the audience seemed to have been satisfied, but my answer, though correct, was far from complete. The idea that there is a genetic link between you and your ancestors is based on the erroneous concept of the blood theory of heredity that I described in chapter 1. Today we know that we inherit our genes from our parents, who inherited their genes from their parents, and so on. But few of us know that the true genetic connection between our grandparents and us is uncertain. The reason for this is that genes are on chromosomes and that we inherit only half the chromosomes of each of our parents. But we have to be more specific. Although we have 23 pairs of chromosomes in our body cells—one pair from our father and the other from our mother—our sex cells have only 23 chromosomes. Which chromosomes of each set go into the sex cell is due to chance by a complex process called *meiosis*, which in great part is responsible for our diversity. Since we have only 46 chromosomes and a tremendous number of ancestors, most of them did not contribute any genetic material to us. This is throwing a wrench into the study of ancestry.

For example, take the case of Maya Lin who skyrocketed

to fame at the age of twenty-one during her senior year at Yale as an undergraduate in architecture. She won the open design competition that resulted in the most influential public monument created since World War II: the National Vietnam Veterans Memorial of 1981-1982 in Washington, D.C. Maya Lin was born in 1959 to culturally-accomplished Chinese immigrants in Athens, Ohio. They were both professors at Ohio University. Her mother, Julia Ming-hui Chang Lin, was a poet and scholar of Chinese literature. Her father, Henri Luan Lin, a renowned studio potter, founded the school's ceramics program and later became dean of its College of Fine Arts. Maya Lin's half sister also became an architect. Obviously, there were very good "genes" for architecture in this family. But, the transmission of genes from one generation to another is not understood by many writers. For example, in the article "The Quiet Power of Maya Lin," in the *New York Review of Books* Fall issue September 29, 2016, Martin Filler wrote:

> Maya Lin's more distant antecedents were no less distinguished, as was revealed in a recent episode of the PBS series *Finding Your Roots*, with Louis Gates Jr., in which the Harvard historian showed her a copy of a scroll that purportedly traces the Lin family's origins directly back to her ninety-ninth great-grandfather in 1092 BCE. Gates also identified one of her maternal great-grandmothers, Ye-Deyi, born in 1855), a gynecologist and pediatrician, who was among China's first female physicians. Lin's self assurance in entering a field dominated by men doubtless owes much to her parents' encouragement, but it is hard not to wonder if genetics had a part.[3]

---

3   Martin Filler. "The Quiet Power of Maya Lin," *The New York,*

I do not doubt that Martin Filler knows a lot about architecture, but he does not know that we are the product of both heredity and environment. Maya Lin is no exception. Her assurance is due to her genes and environment (encouragement) that her parents gave her. As to Professor Henry Louis Gates, he does not know that the likelihood of Maya's ninety-ninth great grandfather giving her any genes are none. It does not make any difference who he was. It is also doubtful that her maternal great-grandmother contributed many genes to her either. However, if both her ancestors had not existed, she would not have been born; both gave to all their descendants the sparkle of life.

DNA and Ancestry

Many people are interested in their ancestry. Since the beginning of the 21st century, they were told that complex DNA screening tests make it possible to determine where in the world their ancestors come from. Two main techniques are used: genetic markers on the Y-chromosome can be mapped to trace paternal ancestry and a similar process can be done on mitochondrial DNA to trace maternal ancestry.[4],

---

Review of Books, Fall issue, September 29, 2016. 52
4 Patterns of genetic diversity are evident in all types of DNA, but they have been most extensively documented in what is called *mitochondrial DNA*. Each of us inherits about half our *nuclear DNA* from our mother and half from our father. This nuclear DNA is the 23 pairs of chromosomes that we have. The mitochondrial DNA resides in different compartments of our cells. Mitochondria are organelles in large numbers in most cells in which the biochemical process of respiration and energy production occur. Each of us inherits our mitochondrial DNA exclusively from our mother. Thus variants, in mitochondrial DNA are inherited purely through the maternal lineage, from a mother to all her children, but trans-

⁵ Both technologies (used by commercial companies) take advantage of the fact that some genetic material is passed down, entirely unchanged, from parent to child—in the case of the Y chromosome passed from father to son, and in the case of mitochondrial DNA passed from mother to child (for both sons and daughters). These technologies, however, have severe limitations.

Mapping Y chromosomes and mitochondrial DNA will only trace two genetic lines on a family tree, where the number of branches doubles with each generation. Continue back in this manner for any number of generations and anyone will be connected to only one ancestor in each generation. The test will *not* connect him or her to any of the other thousands of ancestors[6] to whom he or she is also related. This is a slender thread on which hangs an identity, especially in that tests also have a certain margin for error.

I have said that the DNA companies use two tools. They analyze the DNA of the Y-chromosome of their male customers and the mitochondria DNA of all their customers.

Both DNA have genetic markers. Although both genes and markers are segments of DNA, there is a fundamental functional difference between them. A *gene* is responsible for a protein or other molecule, and these molecules have structural or physiological properties that contribute to the biological functions of an individual. Any alteration in the

---

mitted to the following generation only through her daughters. Although her sons have her mitochondrial DNA, it is a hereditary dead end; they did not pass it to their offspring because sperms have no mitochondria.
5   See appendix 2.
6   The number of ancestors is limited since it grows exponentially, Sooner or later, it will be greater than the number of people on earth, which means we have common ancestors and we are all related.

DNA sequence of a gene may cause a protein to be incorrectly manufactured, resulting generally in a disease condition or sometimes an advantage in the metabolic progress. On the other hand, *genetic markers* are specific locations in the DNA where there is known variability, but this variation has no physical or biological effects on the individual. Genetic markers are used to trace a very limited ancestry.

The difference between a gene and a genetic marker cannot be overemphasized because there are no Jewish genes in spite of the fact that some Israeli geneticists claim that to be true. What they call a Jewish gene is a *genetic marker* on the Y-chromosome believed to be carried by the male descendants of Jewish priest (See chapter 6). To call this marker a gene is to not understand the function of a gene.

The public should also be warned of an important problem that Henry Greely raised in the study of genealogy (a problem that exists since the dawn of humanity), which is false paternity. In the past, we discovered a significant percentage of children were not genetically related to the men who were supposedly their fathers by blood types. Today Y-chromosome testing has also led to these unexpected results. It has been estimated that the rate of false paternity by generation is at least 5 percent. Hence after 10 generations the chance that the eight-times-great grandfather supplied a Y chromosome is just over 30 percent. This is a serious problem for surname searches which use the Y chromosome markers because in many nations men have the surname of their presumed father, but not necessarily his Y chromosome.

While some DNA companies are very clear on the limitations of the products they offer, others are not. Instead of explaining that it is probable that a customer's ancestors came from Russia, for example, they make their clients believe that it is a sure thing. Adverts for DNA screening companies give the impression that your results are unique and that these tests will tell you about your specific personal history, but the very history that you receive could be equal to that of thousands of other people.[7] Conversely, the results from your DNA test could be matched with all sorts of different stories to the one you are given: you cannot look at DNA and read like a book or a map of a journey. I am very critical of telling people they have a certain percentage of ethnicity; for example, my step-daughter, Karen, learned that is 23 percent of English ancestry, or to be more precise 23 percent of Western European ancestry. This seems to be a modern version of the Jig-saw theory of heredity because there is no way to tell where in her body is the percentage of DNA the company is talking about? Again, DNA companies should say the chances that you have English ancestors are x percent. (See Erroneous Concepts of Ancestry Jig-saw Theory figure on page 15.)

Most people overestimate the results of these screening tests. Once, I heard Henry Louis Gates, of Harvard University and host of the PBS show, Faces of America, telling a famous newscaster, that after analysis of her DNA, she was a certain percentage Jewish. He should have said, "You have Jewish ancestors." He should also have emphasized that religion has nothing to do with genetics.[8] The genetic ancestry industry has been criticized not only for overestimating their results, but also for reinforcing the biological concept of race. According to the sociologist Troy Duster the industry is sorely in need of government regulation in regard to claims made

---

[7] I discovered this when I read the result of the DNA search of my adopted niece (a Native American from the Grand Canyon.) What was told about her could apply to any one whose ancestors were of the same tribe as my niece.

[8] Alain F. Corcos. *Who is a Jew? Thoughts of a Biologist* (Tucson, AZ: Wheatmark, 2012).

and accuracy of methods used to pinpoint ancestry, as suggested by the American Society of Human Genetics in 2008.[9]

We should not, however, underestimate the use of DNA in the field of identification of family members, sometimes generations apart. The most recent and interesting case concerns a skeleton found under a parking lot. The University of Leicester confirmed on 4 February 2013 that it was beyond reasonable doubt the one of Richard III, based on a combination of evidence from radiocarbon dating and a comparison of his mitochondrial DNA with two matrilineal descendants of Richard IIII's eldest sister, Ann of York. In 2004, the British historian John Ashdown-Hill had used genealogical research to trace matrilineal descendants of Ann of York. A British-born woman, who immigrated to Canada after the Second World War, Joy Ibsen (nee Brown), was found to be a sixteenth-generation great niece of the king in the same direct maternal direct line. Joy Ibsen's mitochondrial DNA was tested and should be the same as the one of King Richard III. Joy Ibsen died in 2008, but her son, Michael, gave a mouth-swab sample to the research team and his mitochondrial DNA was compared to samples from the human remains at the excavation site. Perfect match: the skeleton was that of King Richard III.

Another example was in 1918 when a wounded woman showed up in a Berlin mental hospital claiming to be Anastasia, the last surviving member of the Russian imperial Romanoff family. Her story, from which she never wavered, engendered an epic controversy that ranged from courtrooms to the silver screen. The mysterious woman married an American, took the name Anna Anderson, and died in 1984. After

---

9   Http:// www. americanscientist.org/bookshelf/pub/race-finished

her death, an amateur historian bought some of Anderson's books. In one was an envelope with some strands of her hair. He took them to Mark Stoneking, a Penn State University genetic anthropologist who would later confirm the identity of Jesse James's remains. Meanwhile an English geneticist had obtained some of Anderson's colon tissue that a hospital had stored after an operation. Both researchers analyze the DNA. "We found that our sequences matched each other," Stoneking recalls, but they did not match the royal family."

So who was Anna Anderson? "One of the private investigators, hired by other Russian nobility, came to the conclusion that she was a Polish woman who had been working in a munitions factory," Stoneking says. There had been an explosion at this factory, which could explain the wounds that gave such credence to her tale of fleeing the Bolsheviks. The English team tracked down a relative of this Polish woman, and indeed her DNA matched Anna Anderson's.

DNA analysis can be quite revealing. Not only paternity might be questioned, but also our oral histories. For example, in Southern Colorado, a group of people trace their ancestry to Spanish settlers from the 1500. Their oral history says they did not have children with the Native Americans. But genetics tells a different story: about 85 percent of their descendants carry mitochondrial DNA of Native American origin. [10]

In brief, biology teaches us that if we know the contribution of our parents, that is half of our genes (or rather forms of genes) come from our father and the other half from our mother, we do not know the exact contribution of our grandparents, even less the contribution of our great-grand par-

---

[10] Mark Scoofs. *What DNA says about human ancestry—and bigotry? Part 3. The Myth of Race. Village Voice.* http://web.mit.edu/ racescience/in-media/-dna-says-about-human

ents. The odds that you have any genes from your famous ancestor six-generations back are very small. Telling us that you are a descendent from Benjamin Franklin might be socially interesting, but means nothing genetically.

5
Ethnicity: A Biological Myth

> *Ethnic groups are categories of human invention, not given by nature. Their boundaries are porous, their existence historically ephemeral. There are the French, but no more Franks; there are the English, but no Saxons; and Navajos, but no Anasa.*[1]
>
> .Jonathan Marks

Ethnic groups are populations of human beings whose members identify with each other on the basis of a real or presumed common genealogy or ancestry. That is, ethnicity is a shared cultural heritage: a sense of history, language, and religion. It refers to a decision people make to depict themselves or others as the bearers of a certain cultural identity. Ethnic differences are not biologically inherited; they are learned.

Genes help us to acquire a culture, but do not determine which culture we acquire. This is why one cannot say, "I am one half-Italian" because Italian is a nationality. Either you

---

1 Jonathan Marks. *What it means to be 98% chimpanzee* (Berkeley CA: University of California Press, 2003), 202

are Italian, or you are not. You cannot say, "I am half Jewish," because Judaism is a religion.[2] Either you are a Jew, or you are not. There are no Italian or Jewish chromosomes, genes, or DNA. We are not born Italian or Jewish. We grow up in an Italian or a Jewish family.[3]

People say "I am one-fourth American Indian," but no one says, "I am half black," because in the United States, as I mentioned before, if your skin is dark you are automatically considered a black person. Black and white are social terms, not biological terms. Skin color is a human trait, not a social trait. It is unfortunate that biology of skin color is taught only to dermatologists, not anybody else, not even to biology majors. I believe that people should know that our differences in skin color, although very visible, are biologically speaking very small. (See Chapter 10)

Although scientific Investigations show that all human groups, however socially defined, are remarkably similar genetically, the general public is still focusing on the few

---

2 Interestingly enough no one says that he or she is one-fourth Protestant or three-fourths Catholic. These are recognized as religious labels, while being Jewish is equally religious-cultural, it is still assumed to be racial.

3 It is hard for writers to abandon ideas or ways of writing when they themselves are pushing for new ones. For example, Robert Wald Sussman writes in *The Myth of Race* page 147, "Boas was born in Minden, Germany. His grandparents were observant Jews, but his parents were not.... Although he [Boas] was not a practicing Jew, Boas was sensitive about his Jewish ancestry and vocally opposed anti Semitism. "

If Boas and his parents were not religious, they were not Jews. As Sussman tells us in another part of his book, Judaism is a religion, not a race. Again on page 167 Sussman contradicted himself when he says Boas was an immigrant Jew. He was an immigrant all right, but he was not, a Jew. He was of Jewish ancestry if that is important.

superficial physical differences among people rather than on their overwhelming similarities. Skin color is so distinctive that, today like in the past, humans are classified with or without color. People still use the term "people of color" to designate those who have a darker skin than Europeans. There are the following problems associated with the use of this term:

1. The term has never been a neutral, objective one. It was invented by early European colonists, who being totally ignorant about different types of cultures, could only identify people according to some superficial and trivial physical differences.

2. It confuses biological groups with social, cultural groups, disparaging the significant influences of learning and social and historical contexts on the formation of group identities.

3. It misleads people to believe that the "people of color" are one group with a similar culture, in contrast with Anglo-American culture. It is proverbial that the so-called people of color are neither culturally nor physically a group.

4. It reinforces the erroneous notion that differences between Blacks and Whites are as great as they could be, and that other groups probably fall in between, according to the criterion of the colors of their skin.

5. It denies that all cultures are involved in exchange relationships, assimilating each other, in particular in modern society in which the transmitting of knowledge and values is no longer based on simple individual contacts.

People have a tendency to believe that ethnicity is inheritable like our physical traits. In our racist world, being a Jew or Muslim is supposedly transferred from parent to child as a hereditary defect, an indelible spot, and a profound reason for not being able to assimilate. But, Judaism and Islam are religions that one can abandon or adopt. The journalist Stephane Charbonnier[4] wrote an open letter to the reader a few days before terrorists assassinated him in January 2015. The title of this letter was, "If you think someone with Muslim parents must be also a Muslim..." Charbonnier was completely aware that ethnicity was not transferred in the genes and that many children of Muslim parents were non-religious. I like to repeat that there are no ethnic chromosomes or genes; yet, orthodox Jews believe that Judaism is transmitted from parents to children biologically. In the past, they believed it was transmitted through the blood; today they think it's transmitted through the genes.[5] In April 1996, the rabbi Yitzhak Ginsburg went so far as to publish an article affirming that in each cell of a Jew –but not in those of non-Jews— there is a substance containing a part of God. According to him, the DNA of the Jews is different from that of the non-Jews. This is why organs can be transferred from Jews to non-Jews, but organs cannot be transferred from non-Jews to Jews.[6] I never

---

4   His nom de plume was Charlie Hebdo. He was the voice of a classic French republicanism willing to grant everything to individuals as individuals and nothing as a group.
5   "Heredity of the Jews" was the first article in the Journal of Genetics in 1911. Redcliffe Nathan Salaman, a famous physician biologist, defends his thesis that according to which, even if the Jews do not constitute a pure race, they form, however, a compact biological entity. Not only can the Jew be recognized by the form of his cranium, his face and his body dimensions, all the characteristics are due to one form of a gene.
6   . www. Le temps.ch/Page/unuid/e64bCb48

heard such a racist interpretation of the well-known biological phenomenon of organ rejection.

There is no doubt that, among Jewish geneticists in the past[7] and Israeli geneticists today,[8] there is a tendency to "prove" that Jews are biologically different, and that they have genes that do not exist in other human groups.[9]

What I say next might be obvious to some readers but it is not obvious to others. There is nothing biological about ethnicity. Like race, it is a human social invention. The Greek word *ethnos* originally meant a number of people living together who generally have the same customs. Associated with ethnic groups are language (Hispanics), religion (Jews),[10] customs (Mennonites), and nationalities (Irish), and also physical characteristics (skin color). Ethnic groups, therefore, implicitly include what has been called "race."[11] One or more (sometimes all) of these distinctions may serve as sources of ethnic divisiveness; any one of them can provoke disdain, discrimination, or violence among groups that do not share traits or customs.

There are still other misconceptions about ethnicity. Ethnic groups are not static entities: they have changed with

---

7 Arthur Mourant, et al, *The Genetics of the Jews* (Oxford: Oxford University Press, 1978
8 Bat-Sheva Bonne-Tamir. "Analysis of Genetic Data on Jewish Populations." *American Journal of Human Genetics*, XXX In 3, 1970 pp.324-330
9 See Harry Ostrrer *Legacy* (Oxford: Oxford University Press, 2012} in particular Chapter 1)
10 Judaism is a religion, yet one can find Jews as a nation or race in official data. For example, one can find a table with the heading "Net immigration of various European nationalities (1820-1930) in Stephen Steinberg's *The Ethnic Myth* (Boston: Beacon Press, 1981),41
11 The terms "racial" and "ethnic" are always associated in the mouth of commentators or under the pen of journalists.

time and so have their names. In Western Europe, there used to be the Gauls and the Celts. Today, there are the French and the Irish. Tomorrow who knows? Ethnic classification can be completely ahistorical. For example, once I read in an old anthropology book that a Cro-Magnon skeleton was that of an old *Frenchman*, (because his bones were discovered in this part of Europe that is now France). A more recent example had to do with thousand-year-old remains discovered beneath the ruins of the medieval town, Dmanisi, in what is now the nation of Georgia. The author of the article calls them Early Georgians.[12]

There are other problems with ethnicity. People are bragging about their ethnicity. But they should not, because they did not choose their parents. They can be proud of their own accomplishments, but they cannot be proud of being white, or black, Irish, French, Jewish, or American Indian descent. They have nothing to do with it. I repeat: they did not choose their parents.

History gives no grounds for ethnic pride: no human group has a monopoly on virtue. If anything, a study of history should inspire humility, rather than pride. People of every origin have committed terrible crimes. Whether one looks at the history of Europe, Africa, America, or Asia, every continent offers examples of inhumanity and genocide from ancient time right through our own day. The last examples are in the Middle East and Africa.

Another problem with ethnicity is that we have a tendency to believe that we inherit the qualities of our ancestors, such as the fighting spirit of the Irish, the thriftiness of the Scotts, or the stubbornness of the Dutch. Qualities like

---

12 Nick Gore. New Find" *National Geographic Magazine*. August 2002, 6.

these might be well achieved in racehorses, fighting bulls, or gamecocks through intensive inbreeding and selection over many generations, but it is inconceivable that they would be transmitted to us, since we generally mate without general preconceived selection.

As there were never pure races, there were never pure ethnic groups. But even if there were in the past, they would have been broken up by "intermarriage." Such marriages are on the rise. As back as 1976, Richard Alba reported gigantic leaps in "intermarriage" of Catholic Americans of different descent including Irish, German, French, Polish, Italian, and Eastern European. In each instance, by the third generation, a substantial majority was marrying outside their group and, in the case of the Irish, over three-quarters were doing so.[13] These figures raise serious questions about the future of ethnic groups that do not speak the language or follow the customs of their ancestors. For example, I have a friend whose name suggests that he is of Italian ancestry, but he does not like tomatoes and does not eat cheese. What kind of Italian is he?

Today the development of swift modes of communication and transportation helps people of different origin to meet and marry each other. In the future, ethnic groups, as we know them today, will disappear or reappear under different names.

The increase in inter-religious marriage is perhaps best illustrated by the patterns of Jews. Today, more Jews (52 percent) marry non-Jews than Jews.[14] Such an increase is not

---

13 Richard Alba. "Social Assimilation among Catholic National Origin Groups.". American Sociological Review 41, 1976, 1030-1046

14 Leonard Dimmerstein. *Anti-Semitism in America* (New York: Oxford University Press, 1994), 241

astonishing in view of the decline of anti-Semitism in the United States. This decline was exhibited by court cases that ended discrimination and ensured freedom of association, as well as changes in residential patterns, which allowed Jews and non-Jews to attend public school together, date, and marry. According to Robert Gordis, an American Rabbi, intermarriage is part of the price that modern Jews must pay for freedom and equality in an open society.[15]

The sociologist Richard Alba has suggested that ethnic distinctions based on European ancestry, once quite prominent in the social landscape, are fading in the background. As a result, other ethnic distinctions appear to be more highlighted.[16] This seems absolutely true when one considers the rate of marriage between white and black Americans. Legal marriages between black and white people, which were rare before the civil war, rose after emancipation, peaked about 1900, and declined until 1940. Beginning sometime after World War II, the number of black-white marriages rose once again, but so slowly that, by 2002, they accounted for only 0.7% of all marriages.

It took until 1967 for the Supreme Court to declare that state laws that prohibited black-white marriages were unconstitutional.[17] The historic legacy of stigma, together, with scant opportunity to meet on an equal footing in offices, schools, and neighborhoods has been the main obstacle for black-white unions. Another has been the repudiation of

---

15   Robert Gordis, *Judaism in a Christian World*. (New York: Mc-Graw Hill, 1966), 186
16   Richard Alba. *Ethnic Identity: The Transformation of White America* (New Haven: Yale University Press, 1990),3
17   Far all the gains in race relations, romance on the big screen between a "black" woman and a "white" man or vice versa remains a taboo.

such unions by black people who view it as an expression of "racial" disloyalty. The majority of blacks and whites also oppose such unions because of social considerations. Even those who approve it in principle find it difficult to advise their sons or daughters to enter into such marriages, knowing the unavoidable social problems that will confront such a couple. The rate of marriage between "whites" and "blacks" is higher in sectors, like academia and the military: this is a hopeful sign for further integration.

To go back to biology, I cannot emphasize enough that marriage between ethnic groups is not only a tool for increasing peace, but is also better for the health of future generations of children. There are genetic diseases that are more prominent in some human groups than in others and children of parents that are from different parts of the world will have a better chance of not inheriting the genes responsible for those genetic diseases. For example, the children of a man whose Jewish ancestors came from Eastern Europe and a woman whose ancestors are from India have no chance of dying of Tay-Sachs disease. The children of a man whose ancestors came from Equatorial Africa and a woman whose ancestors came from Norway have no chance of suffering from sickle-cell anemia. And, although beauty is something that cannot be described, children who are the products of parents coming from different ethnic groups are often very "handsome", quite possibly because they are healthy.

For black Americans, therefore, the problem of ethnicity is far more complex than it is for European Americans, or even Asian Americans. They are physically distinguishable by their skin color and today skin color still matters far more than being Catholic, Jewish, or Hispanic.[18] In chapter 10 I will

---

18   Cornel West. *Race Matters* (Boston: Beacon Press, 1993), 85

show that we know a lot about skin color, far more than we knew twenty years ago. The genetic differences between the skin color of a European and an African are very, very small. When our ancestors came out of Africa, there was natural selection for lighter skin not only in Europe, but in Asia. [19]

---

19   Daniel J. Fairbanks, *Everyone is African* (Amherst, NY: Prometheus Books. 2015) 61-63

6

Genetics and Jewish Identity

> *I find this type of research interesting but not compelling. Frankly there are cultural and religious reasons that are far more significant to me in terms of affirming Jewish solidarity than genetics claims alone.*
>
> David Ellenson [1]

An excellent example of the confusion between ethnicity (in this case, a religion) and biology (genes) is the case of the Jews. Hitler and his cohorts told us, "once a Jew, always a Jew" and the orthodox rabbis tell us, "if you have a Jewish mother you are automatically a Jew.[2] No matter what you do or believe, you remain a Jew." But Judaism is a religion that can be chosen, not an inherited human trait, such as having brown eyes. Therefore, being Jewish has nothing to do with biology, but everything with faith and customs.

In 2008, Shlomo Sand, a professor of history at Tel Aviv University, wrote a controversial book, *The invention of The*

---

1 Rabbi David Ellensin is the [president of the Hebrew College]
2 If your father is Jewish, but if your mother is not, you are not automatically a Jew.

Jewish People. Originally written in Hebrew, it was translated into many languages, including English.[3] The book's thesis is the denial that there is such a thing as the Jewish people, descended from the inhabitants of Biblical Palestine from which they have been scattered, and that they are a nation, which was returned to the land of their ancestors. This undermines one of the principal arguments with which the State of Israel legitimizes itself. The Israeli Declaration of Independence states: "After being forcibly exiled from their Land, the People kept faith throughout their dispersion and never ceased to pray and hope for their return to it."[4] Hence, to justify its existence, the State of Israel has to believe that the Jews of today are descendants of the Jews of yesterday and that there was an exodus.

According to Sand, this obsessively held Zionist-Israeli notion of the Jews as an ethnically identifiable people existing since biblical times is a myth, an invention. And so predictably, this book was greeted with academic,[5] religious, and political fury. No wonder: if there is no Jewish people, but only a Jewish religion; and if the Jewish Diaspora was driven not by forced exile, but rather by the impulse to proselytize, then the Zionist–sponsored "return of the Jewish people" to Palestine in the mid-twentieth century has lost its entire theoretical framework.

However, regardless of the historical facts and their interpretation, many Jews believe that they are a separate "race." This is difficult to understand. After the Holocaust, the concept of a Jewish race should have gone by the way-

---

3 Shlomo Sand. *The Invention of the Jewish People,* (Verso, Paperback Version, 2010).
4 Http: ww.mfa.gov.il mfa./foreignpolicy/peace/guide/pages/declaration %20of%/20establis
5 Shlomo Sand. The invention of the Jewish people, ix

side. Although its emphasis decreased, it did not disappear. Strangely, there was a shift in who believed in the Jews as a race. Many non-Jews look upon Judaism simply as a religion and do not consider Jews as people biologically different from themselves. However, many Jews, including some Jewish geneticists, priding themselves on being God's chosen people, jump to the conclusion that they must be biologically different from non-Jews. They assume that their ancestors have always married Jews. This is just simply wrong. In fact, throughout history, many of them have married non-Jews and many non-Jews became Jews.[6] To form a race, human populations have to be separated.[7] It is common sense that no group of people could possibly remain genetically "pure" while scattered across Europe and the Middle East (and ultimately along the world).

In the English edition of his book, *The Invention of the Jewish People*, the historian Shlomo Sand wrote an afterword that he called, "some replies to my critics." In it, we find these interesting two paragraphs:

> After exhausting all the historical arguments, several critics have seized on genetics. The same people who maintain that the Zionists never referred to a race conclude their argument by evoking a common Jewish gene. Their thinking can be summed up as follows: "We are not a pure race, but we are a race, just the same." In the 1950s there was research in Israel on characteristic Jewish fingerprints, and from the

---

6 Alain F. Corcos. *The Myth of the Jewish Race* (Bethlehem, NJ: Lehigh University Press 2005)
7 Alain F. Corcos *The Myth of Human Races*, 2nd edition, Wheatmark, 2016

1970s, biologists in their laboratories (sometimes also in the USA) have sought a genetic marker common to all Jews. I reviewed in my book their lack of data, the frequent slipperiness of their conclusions. their ethno-nationalist ardor, which is unsupported by any serious scientific findings....

As of today, no study based on anonymous DNA samples has succeeded in identifying a genetic marker specific to Jews, and it is not likely that any study ever will. It is a bitter irony to see the descendants of Holocaust survivors set out to find a biological Jewish identity: Hitler would certainly have been very pleased! And it is all the more repulsive that this kind of research should be conducted in a state that has waged for years a declared policy of "Judaization of the country" in which even today a Jew is not allowed to marry a non-Jew."[8]

Sand was referring to a particular geneticist when he made the remark about Hitler. The man is Harry Ostrer, a medical geneticist, who wrote a book, *Legacy: A genetic History of the Jewish People*. The book is disappointing, because of what the author says and how he says it.[9] According to Ostrer, "the Jews can be said to be a people with a shared genetic legacy, *although not all Jews have the same genes (italics added)*, nor having part of that legacy a requirement for being Jewish."[10] But, all humans have the same genes–otherwise, they would not belong to the same species. What they do not

---

8 Shlomo Sand. *The Invention of the Jewish People* (London: Verso, 2009), 318
9 Harry Oster, *Legacy* (Oxford, Oxford University Press, 2012),
10 Ibid., xviii

have are the same forms of genes, the alleles (see chapter 2). Why does Ostrer write, "if being Jewish were in the blood, then what better way to identify the markers of Jewishness than by studying blood itself?"[11] As a geneticist Ostrer should not use expression "in the blood." On page 213, of his book, Ostrer says, "This period of mass migration of Soviet Jews created mistrust between secular and religious Jews." But, there are no secular Jews, unless one believes that Jews are a race, which, as I said before, is very disputable because Jews marry non-Jews throughout their history. This is recognized by Ostrer himself in his book, *Legacy*, in which he has the tendency to contradict himself. In the book, he argues that Jewishness is biological, not just cultural. Professor Ostrer is wrong because, I repeat again, there are neither specific Jewish chromosomes nor genes. What he is talking about is not a gene, but a genetic marker.

One of these genetic markers is on the Y chromosome and is called the Cohanim chromosome. It is associated with last name Cohen or Cohn, traditionally held among Jews that claim to be descendants of Moses' brother Aaron. What does it mean if someone has this marker? Does it mean that one is a Jew? Obviously not, since men of Christian, Muslim, or any other faith, or no faith at all, can have this marker. The fact that non-Jews carry the Cohanim marker is not astonishing because, historically, there have been a tremendous number of men who have abandoned Judaism and marry non-Jewish women. Their male descendants have inherited the marker. If you have it, it simply means that you might have a Jewish ancestor as far back as eighty generations.

There is no such thing as a Jewish gene, only Jewish an-

---

11   Ibid 119

cestors.¹² There is no genetic link between the Jews of today and the Jews of the past. The only link between them is their religion. Why Jews do not accept this and cling to the idea that they are biologically different from non-Jews is hard to tell. Shlomo Sand suggests that, more likely it has to do with Israeli politics.¹³

The notion that genetics could become a significant tool to determine Jewish identity is incendiary for many, with some Jewish thinkers expressing profound discomfort with the idea. For Ostrer, who is Jewish, being told in print that "Hitler would certainly have been very pleased "for your work" can't be pleasant.¹⁴ But, that is what happened to him in 2010, when he and his colleagues published a study showing that Jews in three different geographical areas had certain collections of forms of genes that made them more biologically similar to one another than they were to non-Jews in the same regions.¹⁵ It is their work and its conclusion, that certain genetic signatures could be used to identify Jews, indicating that Jews share a common biological identity beyond their religious affiliation.¹⁶ But, this is not

---

12   John Dupre. ("What Genes Are and Why There are no Genes for Race" *In Revisiting Race in a Genomic Age*. Barbara A. Koenig, Sandra Soo-Sim Lee & Sarah S. Richardson., eds. New Brunswick, N.J: Rutgers University Press, 2008).
13   Shlomo Sands, The invention of the Jewish People, p.28
14   The quotation about "Hitler being would have been very pleased" is attributed to Sand who told it to a Science Magazine reporter., Sand introduced this quotation in the English translation of his book page 319
15   Gil Atzmon et al. "Abraham's children in the Genome Era: Major Jewish Diaspora Populations Comprise Distinct Genetic Clusters shared Middle Eastern Ancestry." American J. Human Genetics,
June 2010; 80(6):850-859
16   I wrote a similar crack in my essay, "*Who is a Jew? Thoughts of*

a joke. Ostrer in the preface of his book Legacy wrote, "Despite our considerable effort to use the most rigorous of scientific approaches, the discredited race science theories of the Nazis were cited as proof of our misguided behavior. The imputation of equivalence was morally objectionable, because we were not seeking to develop a hierarchy of human groups nor attempting to eliminate individuals on the basis of having "undesirable" genes or traits as the Nazis had."[17] Ostrer missed the point of the debate. His critics do not contest the results of his work, but wonder why he is working so hard to show that Jews are biologically different from non-Jews.

While the study broadly echoed many of the findings of earlier Jewish genetic population studies, which looked at maternal mitochondria DNA or paternally inherited Y-chromosome markers, Ostrer's team studied 160,000 markers on the entire genome. Their study looked at the genetic material of people whose origins lay in seven different Jewish communities: Iranian, Iraqi, Syrian, Italian, Turkish, and Greek. Researchers compared these groups to those of the local non-Jewish populations. They found for more genetic linkage between Jews within each community than to non-Jews from the same areas, and significant linkages between Jews of different communities. The study also demonstrated "distinctive population clusters, each with shared Middle Eastern ancestry."

After reading the article three times, I felt that these thirteen researchers have worked very hard to discover the obvi-

---

a Biologist." (Wheatmark, 2012), 29.

"I cannot prevent myself from thinking how ironical it would be that Adolph Hither had the famous Jewish marker and Benjamin Netanyahu did not ".

17   Harry Ostrer. *Legacy* XV

ous: The descendants of Jews who married Jews will be more genetically related than those of Jews who marred non-Jews. This would be true of any population that limits its choice of marriage partners (In the case of Jews it is religion). It is hard to understand what the conclusions of this work were and how important they were. After all, genetics do not define whether or not an individual should be considered Jewish. There have been people who convert to Judaism—it as been a regular feature of Jewish life. Those who did or do now are no less Jewish than anybody else.

A point that is not made at all in this discussion is that the history of the Jews is that Jews remained Jews. If you select Jews on the basis of their religion to study their genetic relationship, you are doing only half the job. You should study families of "Jews" who abandon the Jewish faith and marry non-Jews. This type of research might be more interesting because, if you believe that Jews are biologically different from non-Jews (and this is a big assumption) such a study might reveal which characteristics are inherited.

The idea that Jews are biologically different from non-Jews is of course not new. It has always been the basis of anti-Semitism. So, why are Jewish scientists so eager to demonstrate that such differences exist? For example, in 1978, Arthur Mourant, a professor of Genetics at Oxford University, wrote *Genetics of the Jews*.[18] Using gene frequency of blood types A, B, and 0, he attempted to prove that Jews were genetically different from non-Jews. However, when the results did not correspond to his hypothesis, Mourant continued his research hoping to have "better" results. Shlomo Sand sug-

---

18 Arthur E. Mourant et al. *The Genetics of the Jews*. (Oxford University Press. 1978).

gested that the work of Arthur Mourant led to the search of the "Jewish" gene for the same reasons.[19]

Tracing genetic ancestry is hard and causes more problems than it solves. For example, confusion looms when genetic markers conflict with other markers of group membership, such as shared culture or historical narrative. The philosophical question then arises: does it make you any more English, or Sioux, or Jewish if your identity has been confirmed by a genetic marker? [20]

We see confusion in papers and other writings, of what science reveals in the field of genealogy. Let us go back to the markers on the Y-chromosome. In an article from the Los Angeles Times, entitled *DNA Clears the Fog over Latino Links To Judaism in New Mexico*, one can read the following:

> As a boy, Father William Sanchez sensed he was different. His family spun tops on Christmas, shunned pork and whispered of a past in medieval Spain. If anyone knew the secret, they weren't telling, and Sanchez stopped asking. Then three years ago, after watching a program on genealogy, Sanchez sent for a DNA kit that could help track a person's background through genetic foot printing. He soon got a call from Bennet Greenspan, owner of the Houston-based company.[21]

And this is where *Father* Sanchez's story becomes inter-

---

19   Shlomo Sand. *How the Jewish people was invented?* (Verso, 2009) 274
20   S.A. Persectives: Racing to Conclusions. *Scientific American* (December 2003)
21   David Kelley. DNA clears the fog over Latino links to Judaism in New Mexico. *Los Angeles Times,* December 5 2004.

esting. Greenspan asked him if he knew that he had Jewish ancestry. Then, he was told that he was a Cohanim, a member of the priestly class descended from Aaron, the brother of Moses. It is highly possible that the Sanchez family came from Spain and was at one time Jewish. But, of course, as I said before, one should make a distinction between having a Jewish ancestor and being Jewish. After all, Sanchez is a Catholic priest. However, no places in the article can one find that its author has understood that we should make the above distinction. My point is that it is the duty of journalists to understand what they write about, because they are the communicators between scientists and the public.

It might be more interesting to apply modern genetics to a human population whose origin is completely unknown, for example the Melungeons, an Appalachian people in Hancock County Tennessee. Some have dark skin. The county is isolated, beautiful, and poor. Some of its residents resorted to the usual shifts of penury, moon shining, and the like. Those exigencies, combined with a reputation of bushwhacking during the Civil War—and above all, the enduring queasiness about miscegenation—turned the Melungeons to their neighbors, into renegades and bogymen.

They are said to be the progy of Phoenicians who fled the Roman sacking of Carthage, or pre-Columbian Turkish explorers. They may descend from wayward conquistadores, or from a doomed colony established on Roanoke Island by Sir Walter Raleigh, or from Moorish galley slaves abandoned there by shipwrecked pirates or by Madoc, a 12th century Welsh explorer, or even they are a lost tribe of Israel.[22] However, one of the most widespread beliefs is that they are off-

---

22 Anonymous" Down in the valley. Up on the ridge. *The Economist*, August 27th, 2016

spring of Portuguese mariners who arrived in early colonial times. Recently DNA tests were applied, but they were inconclusive, possibly because so many Americans are descendants of immigrants from all over the world and that marriages, especially on the colonial frontier, were at random.

7

Each of Us Is Biologically Unique

What was the most extraordinary adventure in your life? Whatever your answer, You are almost certain to be wrong. For the most remarkable series of events that could possibly have befallen you took place before you were born. In fact it was virtually a miracle that you were born at all!

Abram Scheinfeld[1]

When I was a teenager, I read in French the novel, *The Life and Opinions of Tristram Shandy, Gentleman,* by Laurence Sterne. It contained a discussion I will always remember. The author wanted to stress, like Abram Scheinfeld in the epigraph above, that the chances of having come into existence were very, very small. To illustrate his case, he declared that a few minutes before his conception, his mother asked his father if he had wound up the clock. He remarked that if his mother had not done so, he would not have been around to write his novel. His parents would have a child, but it would not have

---

[1] Abram Scheinfeld. *Your Heredity and Environment.* (New York: Lippincott Company, 1965) 27.

been him. Laurence Sterne knew his biology. To explain further, the egg that produced him would have been the same in any case, since a woman generally has only one egg per month, but men produce millions of sperm every day. The chances that a particular sperm fertilizes an egg are infinitely small. Since each person is the result of the fertilization of one specific egg by one specific sperm, the conception of our novelist was a pure accident, and so is conception of each of us.

You are unique. Consider what had to happen for you to come into existence. First *you*—that very person who is *you* and no one else in this universe—can only be the child of two specific parents out of all the men and women on the earth. It was an amazing coincidence that your parents came together. But taking that for granted, what were the chances of their having you as a child? The answer is that those chances are very small, very small indeed because the minimum number of children of different kinds could have is in the millions (to be exact 2^n, n being 23, or 8,388,608). Just consider the number of chromosome combinations any two parents can produce in their eggs or sperms. For what every parent gives to a child are replicas of just half of his or her chromosomes—one representative of every pair taken at random. Although the chances that your parents had you as a child are smaller than 1/8,388,608 (other factors decrease the chances) there are small enough to make you realize how lucky (or unlucky) you are to be alive.

Each of us is unique for many reasons, which were historically discovered as follows: (1) although we have the same number of genes, the forms of these genes may differ from one person to another due to the process of mutation; (2) the biological process of meiosis contributes to this diversity, which occurs during the formation of eggs and sperms; (3) the expression of our genes is influenced by other genes, which can also mutate; (4) the expression of genes can be modified by the environment; (5) the fact that we grow and live in different environments helps shape us as individuals.²

The Process of Mutation

Mutations are sudden heritable changes in the structure of the genetic material. For centuries, plant and animal breeders have observed occasional *sports* –plants or animal traits that seem to occur by chance and are transmitted to the next generation. For example, Herefords have been for a long time the most popular ranch cattle in the United States and Canada. When in 1889 a hornless calf was born on a Kansas farm, stockbreeders recognized the advantage of ranch cattle without horns. From this Kansas calf have been bred all the present "polled" Herefords." Hornless was a useful mutation. Sometimes a peach tree with fuzzy peaches will bear a non-fuzzy peach. If the seed of that peach is planted, it will give rise to a tree bearing only non-fuzzy peaches. Non-fuzziness was due to a mutation of a gene.³

In 1927, H.J. Mulller studying the fruit fly, and L.J. Stadler,

---

2 Many parents think of their children as little duplicates of themselves, but their children have different complements of genes and are developing in a different environment. This misunderstanding might be very cruel as the following story shows. Years ago, a black woman I knew had suffered so much discrimination that she did not want her five-year-old girl to suffer as much as she did. In order to ensure that, she intentionally shot her child through the brain. Fortunately, or unfortunately, the little girl survived. I, like many others, was devastated because the little girl had been a joy to everyone who knew her. She was smiling, dancing, and full of life—not like her mother, who thought that her child was an image of herself.
3 For two years of his life the author was a peach breeder.

studying corn, independently discovered the mutagenic effects of ionizing radiation. This discovery led to further investigations, which showed that several kinds of radiation, including ionizing X-rays and ultraviolet light, could induce mutations in genes. While significant in demonstrating that the mutability of genes could be enhanced by environmental alterations, radiation experiments were not too helpful in elucidating the physical and chemical basis of mutation, because they produced translocations[4] and inversions.[5] Ultraviolet light, on the other hand, causes few chromosomal rearrangements and losses, and more "point" mutations[6] (see fig 1).

It took years to exactly understand the real structure of a point mutation.

Biologists today can create mutations at will in any organism, using X-rays or mutagenic agents. Natural or induced mutations are changes that occur in our DNA sequence, either due to mistakes when DNA is copied or as a result of environmental factors such as UV light and cigarette smoke.

Over a lifetime, our DNA can undergo mutations in the sequence of bases A, C, G, and T (see chapter 2). This results in changes in the proteins that have been made. This can be bad or good. Mutation can disrupt normal gene activity and cause diseases, like cancer, but on the other hand, mutations contribute to genetic variation within species and often permit organisms to survive a bad environment, in other words to evolve.

---

4   Transfer of part of a chromosome into a different part of a homologous, or into a non-homologous chromosome
5   Reversal of a part of a chromosome so that genes within that part lie in inverse order.
6   Point mutations refer to mutations inside individual genes.

Figure 1 *An illustration to show an example of a DNA (point) mutation.*

> Original sequence: TAAC[T]GCAGGT
> Point Mutation: TAAC[C]GCAGGT

Meiosis: The Secret of Human Diversity

Meiosis is the biological phenomenon that is the basis of the source for individual human diversity as it is for all organisms. Meiosis is very complex, but you do need to know all the details to understand its incredible importance.

The fact that genes can have different forms (alleles) created by mutation is one of the fundamental reasons for our diversity. Each of our cells carries between twenty-five and thirty thousand genes, each of which might be a different form. However, mutation by itself would not be that important if it were not accompanied by meiosis, the process that creates sex cells.

Each of us originates from one cell—one egg fertilized by one sperm. This original cell divides into two new cells, which then divide into two new cells, and so forth. Within the original cell, chromosomes duplicate, and each of the two new cells receives forty-six chromosomes (twenty-three pairs) identical to those contained in the original cell. The process by which a cell reproduces into two identical cells is called *mitosis*. Mitosis is involved in the production of new cells, the growth of new tissue, or the repair of existing cells. Except for occasional mutations, our body cells all have the same DNA.

We understand why meiosis exists: there should be a reduction in the number of chromosomes every generation, because otherwise the number of chromosomes would double every generation (46, 92, and 184). But why such a complicated process? Meiosis is very difficult to understand and even harder to teach. Nature could have devised a simpler process, but it did not. Sex cells are the product of two consecutive, special divisions in which the forty-six chromosomes, after they have duplicated, are distributed so that each of the four new cells receives only one chromosome from each pair, twenty-three chromosomes in all. One of these is a sex chromosome; the other twenty-two are called autosomal chromosomes. Each egg produced by a woman has twenty-two autosomal chromosomes and one X chromosome. Each sperm also has the twenty-two autosomal chromosomes but there are two types of sperm, one with an X chromosome and one with a Y chromosome. At fertilization, the union of an egg and sperm forms a new cell in which the characteristic number of chromosomes (forty-six) is restored again. The embryo will generally be male if the egg is fertilized by a Y sperm, and female if fertilized by an X sperm. Each sex cell is genetically different from all others.

Fertilization of the egg by the sperm results in the formation of a new individual (known as a zygote), who has each kind of gene duplicated, derived respectively from the mother and father. The form of the gene may be alike and geneticists write the genotype of the child (A1A1), in which case the individual is said to be homozygous for the gene in question; or the form of the gene from each parent pair may be somewhat different and geneticists write the genotype of the child as A1A2, in which situation the individual is said to be heterozygous. A1 and A2 are different forms of the same gene and are called alleles. The gene for earwax, for example, has two alleles: one for dry wax, the other for wet wax.

When a heterozygous father (A1A2) forms sex cells, half of these sex cells (gametes) receive the allele A1 and the other half the allele A2. If the mother is also heterozygous, for the alleles A1 and A2, three types of children can be produced: A1A1, A1A2, and A2A2. Since we believe that human beings have around 30,000 genes, an individual has the potential to produce at least 2^n (n being 30,000). Therefore, it is obvious that a couple can produce children that are very diverse genetically. Yet this diversity is even greater than I just explained, because more than two alleles may exist of the same gene (A1, A2, A3, A4 etc), increasing the individual's diversity.

Our genetic individuality depends in great part on which chromosomes we inherit. Although each of us received half of our chromosomes from our father and half from our mother, each of our parents also received half of their chromosomes from one parent and the other half from the other parent. Our father received twenty-three chromosomes from his father (paternal chromosomes) and twenty-three chromosomes from his mother (maternal chromosomes). However, because each of his sperm contains only one chromosome of each pair (which is determined at random), the sperm that fertilized the egg of my mother to produce me did not receive the same combination of chromosomes as the sperm that fertilized the egg of my mother to produce my brother. The same can be said for the eggs of my mother. Any sex cell from my parents could have received either fifteen paternal chromosomes and eight maternal chromosomes or thirteen maternal chromosomes and ten paternal chromosomes to name just a couple of possibilities.

Since we have twenty-three pairs of chromosomes, there

are 2^n (n being 23) or 8,368,608 possible combinations for a person's gametes, assuming that two forms of the same gene govern one trait and no other factors influence the process. The chances that you received the same combination of chromosomes from your parents as your sibling received is the product of 8,368,608 by 8,368,608, which, for practical purposes, is infinity.[72] This means that two children of any couple of parents, unless they are identical twins, have no chances of having the exact same chromosomes. We can say that any human being is the result of a unique combination of chromosomes (and consequently forms of genes) and he or she is not a reproduction of anyone, but a definite, unique creation, due to chance. It is not possible to predict which genes will be present in a child because there is no way to know which sperm among the millions present in a sexual act will fertilize the egg, nor which genes will be present in the egg that is fertilized. To understand the extent of our diversity, think about this: the number of sperms that have been formed by all the men who have ever lived is far less than the number of possible genetic forms of sperm.

And still our individual diversity is greater than I described above. In my discussion, I have *not* introduced the important fact that chromosomes can break during meiosis. If two genes are located in the same chromosome, and the chromosome remains intact in inheritance, the two genes should remain together (linked) in all cases. This is not what happens. Most of the time, chromosomes break, and part of one chromosome joins with part of its homologous chromosome. In other words, these two chromosomes, one from each

---

7   2. Butler, John M. *Forensic DNA Typing, Second Edition: Biology, Technology, and Genetics of STR Markers*. Second ed. Cambridge: Academic Press, 2005.

parent, randomly exchange sections of DNA, like two sets of cards being crudely cut and mixed. This process, called crossing over, permits a much greater variety in the offspring than would be possible if each chromosome remains intact. Because segments of chromosomes are scrambled every generation, no chromosome passed from one generation to the next remains the same. I call this "chromosome individuality."

Genes Are Influenced By Other Genes

Each person has the same set of genes, about 30,000 in all. The differences between people come from the slight variation of these genes, which are segments of DNA. However, it has been known for a long time that the amount of DNA in our chromosomes was far greater than the amount of DNA in our genes. We did not know why. We thought that a lot of DNA had no function. Geneticists called it Junk DNA, and it was a chock for them to discover that some of this DNA was genes influencing the genes that produce our proteins, hormones, and other products. These newly discovered genes could mutate increasing our individuality.

That we are biologically unique is very important, because this notion completely undermines the assumption that groups are more important than individuals.

I already mentioned that DNA became a very useful forensic tool because each of us has a unique sequence of DNA.[8] This fact well known today, as every viewer of the television series *Cold Case* is reminded in every episode. Since the year

---

8   Although much of the DNA among different individuals is the same, there are regions within the DNA that vary between individuals, which makes us unique.

2000, we know that the human genome contains about 3 billion DNA units, or base pairs, which are letters of the genetic code. The difference in DNA, between two individual human beings, amounts on average to at most 0.05 percent, so there are roughly 3 million base pairs between my neighbor and me. This seems a lot, but it is not when you consider that the difference between a human being and a chimpanzee is about 15 times as much. That equates to 45 million base pairs, the equivalent of the number of letters in the Bible.

Recently, with the mapping of our DNA, we have discovered that our individuality is greater than we thought. Stephen Scherer, a medical geneticist, found more than four million differences between the chromosomes someone had inherited from his father and his mother. Extrapolating from this finding, he suggested that the amount of variation between humans was not 0.01 percent, as the Human Genome Project had estimated, but more like 0.05 percent.[9] 0.01 or 0.05 percent Whatever is the right percentage, it is important to remember that there is more variation between individuals, that between groups of human beings as Richard Lewontin explained a long time ago.[10]

I have given many biological reasons to accept the fact that each of us is unique. We are each the product of a unique heredity and a unique environment. There has never been anyone identical to us. Variety of the human population is astonishing—even to a biologist. Among seven million members of our species, we find no duplicates. Literally each person is biologically unique and declares this fact not only in

---

9   Jon Cohen "The Human Genome, a Decade Later." *MIT Technology Review* 114.1 (2010): 40. Print.
10 Richard C. Lewontin. "The Apportionment of Human Diversity" *Evolutionary Biology* 6: 1972, 385

his or her obvious physical features,[11] but also in chemical constitution and behavior.[12]

Biology leads us to one very important philosophical conclusion: we are unique and should be judged individually, regardless of what group someone put you in. However, biology will never be able to answer the philosophical question of why we are so unique?

---

11   Sometimes we found some people who look like us. I have known two who really look like me. This type of physical similarity has been used in wars to confuse the enemy. General Eisenhower and General Montgomery had one each who played their role very well.

12   A very recent discovery is that each of our brains is unique. Our differences in the way we think, learn, and behave are due to jumping genes, which are sequences of DNA. Jumping genes have virtually been in all species. They pass copies of themselves into other parts of the genome (the full set of DNA) and alter the functioning of the affected cell, making it behave differently from an otherwise identical cell right next to it. These jumping genes explain why identical twins, which have the same DNA and often the same environment, may act differently.

8

Nature and Nurture: You Cannot Have One Without the Other

Nature versus nurture is dead. Long live nature via nurture.

I have often heard the expression that an individual has inherited his eye color from his father and his disposition from his mother. Biologically this is obviously incorrect, since heredity is transmitted through sex cells, which have no eyes in them and no definable disposition. It is not the eye color or other "traits" that are inherited, but something within the cells that determine those traits in concert with the environment in which the individual lives. This "something" is the genes, which are part of the DNA.

The modern concept of the gene is very far from the Mendelian one. Since the time of Mendel, we have discovered that dominant and recessive genes are exceptions. Most genes express themselves differently within a large range of environments. Seen from our perspective, the early geneticists had it easy. For every trait (phenotype) they invented a gene that could have two forms, named alleles. Here is an example: among Mendel's peas, there were both tall and dwarf types. Mendel designated tallness by T, and dwarfness by t. If the pea was tall, its genotype (genetic make-up) could be TT or Tt, with T being the dominant gene. If the pea was a dwarf plant, its genotype was tt. Those traits were expressed regardless of the environment in which the peas were growing.

There are many traits in plants and animals (including human beings) whose inheritance could be explained by the Mendelian laws of inheritance. But others are very complicated traits, such as behavior and intelligence, which, so far, we know nothing about the inheritance of. This did not prevent former researchers from inventing genes for each trait they were working with; no matter how complicated those traits were expressed. Genes were invented for love of the sea and sea sickness and even for stealing and killing.

Of all human traits, only blood types are largely determined by genes and do not seem to be influenced by the environment. A person's blood type does not change throughout his or her lifetime. This property has helped us to safely transfuse blood from one person to another, to determine paternity and maternity in families, and determine criminality before DNA analysis took over that function.

It took a while for geneticists to discover that genes and genotypes express themselves differently in different environments. For example, let us consider a type of variation in corn kernels. The outer tissue of the grain, technically known as the pericarp, shows several variations in color. A certain dominant gene, which we shall symbolize by R, produces red pericarp. But the red color will develop only on kernels that have been exposed to sunlight before and during maturity. In other words, the husks must be removed from kernels of genotype RR or Rr in order to obtain red pericarp. Is the variation in color due to heredity or environment? The red color can appear only in an ear of genotype RR or Rr, but the red color could not have developed in the absence of direct sunlight during development. An ear of genotype rr would show no red pericarp even if a portion of the husk had been removed and some of the kernels were exposed to direct sunlight. In other words, both the proper heredity and environment are necessary to produce that variation within an ear.

The following is an illustration involving animals. The fat of rabbits may be either white or yellow. A dominant gene, which we shall represent by Y, results in the production of white fat, while its recessive mate y results in yellow fat. The Y gene results in the presence of an enzyme in the rabbit liver, whose function is to break down the yellow pigments present in foods eaten by rabbits. In the absence of this enzyme, the yellow pigments are synthesized into yellow fat. But some foods eaten by rabbits contain no yellow pigment. A rabbit fed solely on a ration of potatoes and mash develops white fat, regardless of genotype. The color of the fat of a rabbit fed carrots and mash, however, is determined by his genotype, as carrots contain an abundance of yellow pigment. In order to have yellow fat, a rabbit must be a genotype yy, and must be fed a diet containing at least some yellow pigment. Neither heredity, nor environment can be said to be more important in the determination of rabbit fat color, as both are essential.

Another example from rabbits will also show that the environment changes the expression of genes. The Himalayan strain differs from other hair color varieties by a single allele of the color series. Animals, that have two of these alleles –one from each parent, show a different hair pigment response, depending upon the temperature of their skin. These rabbits have albino hair on the portion of their skin with a temperature of 34C, but develop a melanic pigmentation in the hair on any portion of the skin subjected to temperature below 34C.

The expression of genes can be affected by such factors as temperature, weather, and others. It is possible to create mutations in bacteriophages (viruses that attack and kill bacteria) that cannot live at 37C, but do at 25C.[1] Since 1963, temperature sensitive mutants have been found in many species of plants and animals. The *Bicylus anyana* butterfly grows up to be colorful if it is born in the rainy season, but gray if it is born in the dry season.

Here is a curious case among human beings, which illustrates that you cannot separate heredity and environment in the expression of genes. A twenty-year old man who had an identical twin was struck in the right testicle with a board. Soon afterwards, cancer developed in the testicle and caused his death. Naturally, people concluded that the injury had caused the cancer, but, ten years later, his twin brother, who had not had any injury, developed the same cancer and died. Apparently both boys had inherited a tendency for cancer of the testes, but the onset had been hastened in one, due to injury.

I cannot stress enough that any trait, normal or abnormal, is due to both heredity and environment. I have a cat with a deformed tail, which cannot be straightened out. I suspect that this deformation is due to an accident that happened when the cat was in the womb, and that her mother had a big litter. But of course, to have a tail in the first place, the cat needed the genes to grow one.

Genes aren't destiny–especially when it comes to mental

---

1 Alain F. Corcos and Eduardo Orias "Temperature-sensitive mutants of the Salmonella bacteriophage P22." *Genetics* 48: 806

illness. In 2010, it was reported that a particular gene might increase the risk of depression, but only in combination with an added non-genetic factor–a stressful life event. People with one form of a protein that ferries serotonin, a mood related neurotransmitter, are especially prone to depression when faced with a traumatic event, such as being diagnosed with a medical illness or being a victim of childhood abuse. The version of the gene that these individuals carry prevents nerve cells in the brain from reabsorbing serotonin, which leads to a feeling of sadness and a negative mood making it harder for them to emotionally recover from a crisis.

In spite of such kinds of examples and although the idea that both heredity and environment play a role is a simple one, there is still an ongoing debate over whether nature or nurture is more important not only among the public at large, but sometimes among scientists who should know better. For example, in 2005 in France, an education report appeared, recommending the screening and medical treatment of young children who displayed abnormal behavior, such as hitting their schoolmates, not being able to stay put, or lying all the time. In 2006, such reports were heavily criticized because they were based on the assumption that once a child is born it cannot change. Such determinism was rejected, but, strangely, the critics seemed to believe that the environment only plays a role after birth.[2]

In 2009, a neurobiologist at McGill University in Montreal was working on the trait of perfect pitch. He seemed to be astonished that it was impossible to distinguish the influence of the genes and the environment on this particular human trait. Why should it be different from any other trait? Yet, he

---

2   Sylvane Giampino and Catherine Vidal. *Nos enfants sous haute surveiliance* (Paris: Alabin-Michel, 2002)

tells us that some children evidently have a genetic disposition for a perfect pitch, but only if they are taught the musical notes before they are six years old. No adult has this ability if he or she has not been exposed to this training before the age of six.[3] This is a perfect example of interaction of genes and environment. The genes express themselves only in a particular time of the development of the human brain.

Even an old man like me still has the same DNA—the same genotype—as when I was born, but my phenotype has completely changed. I have baby pictures of myself, and it is hard to believe that the baby is now me. Hence, the process of aging is a perfect example of a timing development in human beings. Some traits express themselves with a relatively short period of life span, such as sexuality, while others may span over a wide range of ages, such as hair color.

Another example of timing is the pronunciation of a foreign language. It is now well known that a child who immigrates in the United States before he or she is seven, or even ten years old, will speak English like another child born in the United States.[4] But an adult, like me, who came in America at the age of 22, cannot. Biologists have found the reason for this ability; Children use one particular part of their brain; adults use two parts. I know two brothers who came to the U.S. from Great Britain at the age of 10 and 12. The oldest still speaks with a faint English accent; the other speaks with an American accent.

For decades, plant and animal breeders attempted to partition, in some quantitative way, the contribution made to

---

[3] Mozart et la popo stimulent le cerveau. *Science et* Avenir. October 2009, 76

[4] Unfortunately for the author he came too late in this country, He speaks English with a very strange accent.

variation by differences in biological inheritance, as opposed to environmental differences. Developing sophisticated statistical methods, they were generally successful in developing better animal breeds and plant varieties. However, we cannot, for ethical reasons, employ these methods with human beings, since they involved control breeding.

Using other approaches, such as studies on identical twins (who have the same DNA), we have attempted to create some idea of the contribution of the genes and the environment. Statisticians call this the "heritability value" of a trait. Unfortunately, this type of study is very difficult to do because identical twins, that had very different environments, are hard to find. But, even if we could calculate the true value of heritability of a human trait, it would not mean much. In a recent article, Richard Lewontin, expresses that idea:

> It is, for example, all very well to say that genetic variation is responsible for 76 percent of the observed variation in adult height among American women while the remaining 24 percent is a consequence of differences in nutrition. The implication is that if all variation in nutrition were abolished then 24 percent of the observed height variation among individuals in the population in the next generation would disappear. To say, however, that 76 percent of Evelyn Fox Keller's height was caused by her genes and 24 percent by her nutrition does not make sense. The nonsensical implication of trying to partition the causes of her individual height would be that if she never ate anything she would still be three quarters as tall as she is.[5]

---

5   Richard Lewontin. It is even less in your genes, *The New York*

Yet, in 2010, Time magazine published an article on longevity in which it is said: "only about 30% of aging is genetically based, which means that the majority of other variables are in our hands."[6] It does not mean anything of the sort, because with humans, you cannot experimentally separate the action of the genes and the environment. However, we can improve the environment, which seems to be the reason we live longer, whatever our genes for longevity are.

The impact of the environment on the expression of our genes starts from the day of our conception and lasts until our death. This is very important because it explains why there are no two identical individuals, not even in the case of identical twins, who have the same DNA, but do not grow in identical environments. For the same reason, there are no such things as true clones. This scientific fact was finally accepted in the media about cloning pets. A dog owner spent $155,000 to clone his late yellow Labrador because he wrongly believed that he had the opportunity to have the exact same dog. No one told him that his new dog could not be his late dog, because it will not live in the same environment. His new dog might not even look the same or have the same behavior. Here are the reasons

When Ian Wilmut and his colleagues at the Roslin Institute. University of Edinburgh (Scotland), announced the successful cloning of a sheep from the mammary cells of an adult female, the world reacted with intense emotion and misunderstanding. The arrival of Dolly made it clear that human beings would soon have to face the possibility of human cloning and it has been this idea, far more than the reality of animal cloning that had caused public anxiety. However, this

---

review of Books. May 26, 2001, 23
6   The Science of living longer *Time*. February 22, 2010, 61

fear was based on misunderstanding the biological nature of Dolly. Dolly was not a clone.

Although she had the same DNA as the one of the mammary cell, everything else came from her surrogate "mother." Genes express themselves differently in different environments and the environment in which Dolly grew as an embryo was different than the one of the sheep from which the biologists took the initial mammary cell. That alone tells us that Dolly was not a clone.

Biologists remind us that human clones occur in nature: we call them identical twins. But this is a misnomer. Identical twins are not identical, in the fact that they do not have identical fingerprints and they have their own personalities. If plant clones exist—they can be propagated asexually—animal "clones" do not exist in nature or in the lab. Any "clone," as Dolly will end up being different from the original, even more different than identical twins, given that such "clones" will also be born into different generations. Identical twins are far better clones than Dolly and her genetical "mother" because they share the same womb and the same time and culture (even if they fall into the rare category of siblings separated at birth and rose, unbeknownst to each other, in distant families of different social classes.[7] For all these reasons our dog-lover could not get back his old dog.

Belief that either heredity or environment alone is all-powerful has led to catastrophes. In the 1960's, under the influence of Secretary of Agriculture L.T. Lysenko, the Soviet Union believed the environment could directly alter heredity. The result was that for many years, progress in agricul-

---

7 Stephen Jay Gould. Dolly's Fashion and Louis's Passion *Natural History* vol 106, No 5, June 1997 pages 18-23

tural research in that country was completely stopped.[8] On the other hand, belief in the total power of heredity–as it was believed under Hitler–has been responsible for terrible social evils, such as institutional racism and genocide. The perceived separation of hereditary and environmental traits has led to a total misunderstanding of ourselves in many fields, leaving us with little hope of progress in education and social conditions.

The human body develops and grows responsively to conditions of life. Among these conditions is culture, which includes knowledge, belief, art, language, morals, law, customs, and other capabilities and habits acquired by man as a member of society. The key word in this definition is 'acquired' (i.e., not biologically inherited). It is far from unheard of for someone to acquire a culture completely different from their biological parents.

Culture is wholly acquired through imitation, training, and learning from other human beings. But let us be clear–when we use the expression "biologically inherited," we do not mean something simply handed down readily from our ancestors. On the contrary, as I said before, the characteristics of an organism develop through the long and complex interaction between heredity and environment. Biological heredity may make acquisitions and transmissions of culture (or some of its aspects) more easy or difficult to acquire, but it does not determine just what is acquired or transmitted.

Racists speak and write about heredity as though it was some implacable destiny, the decisions of which cannot be

---

8   In 1967, the Soviet Union finally sent a scientific mission at the annual meeting of World Geneticists in Berkeley, California. The Soviet Union delegates were well received and learned about the progress in DNA research.

appealed. Such a hopeless and defeatist view is not necessary if we realize that heredity determines the response of the person to the environment and that the environment can be changed.

The best examples to show how we can change the results of heredity are in the medical field. Some people cannot retain and utilize sugars, which they get in their foods; instead, the sugars are expelled in the urine. These people are diabetics, and this disease is hereditary. However, a long time ago, it was discovered that injections of insulin promptly relieve the attacks of diabetes. The injections do not cure the illness, and the attacks recur unless injections are made at proper intervals. However, if given regular injections, diabetics lead normal and happy lives. You may not suspect that your friend is a diabetic if he or she creates an artificial environment containing insulin.

The same is true for behavior. Good or bad behavior is not simply inherited. It also depends on the environment in which the individual develops. Fortunately, a change in the environment can transform the nature of a person. It is harder to do this than it is to change a medical condition, but it is possible. To believe that we cannot change our behavior is to deny our humanity.

A while ago, there was still a misconception that genes would explain all human variation. And if we knew all our genes we would be a long way to understanding ourselves. The Human Genome Package was in part sold to the public as a way to find solutions to many of our social, as well as our medical problems. There is no doubt that the human genome is useful in medicine, but I doubt that it will help sociologists in their understanding and remedying of society where environment plays such a role.

How genes actually influence human behavior, and how

human behavior influences genes, might not be known for a long time. However, we do not need to know all the details of the interaction between heredity and environment to be sure that our modern understanding of this biological phenomenon goes in the complete opposite direction of the social Darwinists and racists–those who have urged all along that the existing stratification of the social classes is a reflection of their "native" abilities. On the contrary, being adaptive and educatable are very important qualities of mankind that have permitted not only its survival, but its partial conquest of nature. We have been able to profit from experience, adjusting our behavior to the requirements and expectations of our surroundings. We know enough to contradict one of the tenets of racism, that the development and thinking of an individual is the product of only his or her genes. The more we know about genes, the more we know that they express themselves differently in different environments.

I would like to finish this chapter with these words of the French Nobel Prize winner, Francois Jacob:

> Today biology tells us that any organism, from the most humble to the proudest, is the result of a genetic program that contains all the necessary instructions to its development. But this program cannot unroll only with a permanent contribution of matter, energy, and information. The individual is the culmination of a constant interaction between its program and the milieu in which it lives.
>
> In the very simple organisms the genetic program is rigid; but the evolution of the living world was characterized by a progressive opening of the program: among the superior animals, man in particular, a part of the program does not define precise

properties, but only potentialities, capacities, which develop only according to the individual adventure.

There was a while ago a misconception that genes would explain all human variation. And if we knew all our genes we would be a long way to understand ourselves. The Human Genome Package is sold to the public as a way to find solutions to many of our social, as well as our medical problems. There is no doubt that the human genome is useful in medicine, but I doubt that it will help sociologists in their understanding and remedying of society where environment plays such a role.

By describing the genome as a language that still needs to be deciphered we perpetuate a widespread misconception of the role of genes in the development and functioning of biological organisms Ever since the cracking of the genetic code more than half a century ago, it has been clear that DNA sequences (genes) serve as templates for the manufacture (indirectly) via messenger RNA of the structural proteins and enzymes that constitute the raw materials from which living cells construct and replenish themselves. Not only the majority of DNA does not code for any proteins that are actually synthesized, but all except a minuscule part of everybody's genome will turn out to be exactly the same, since identical biochemical mechanisms for making the same range of cell types are present in all individuals, including most species.

9

Two Revolutions in Biology

Race has no basis in the genetic code
               J. Craig Venter

At the dawn of the twenty-first century two advances in genetics shook the world of biologists; the first being the new science of epigenetics. Epigenetics is the study of cellular and physiological variations that are caused by environmental factors that switch genes on and off and affect how cells read genes instead of being caused by changes in the DNA sequence. It has shown that the environment plays a greater role in the development of organisms than we thought previously by influencing the expression of genes. Environmental factors like stress and nutrition can cause the divergence of genes by changing their behavior. Epigenetics is involved in autism and other neurodevopmental diseases.[1]

  The geneticist E. Miller tells us:

---

1 i. P. Miller "A Thing or Two about Twins" *National Geographic* 221, 38-65

"Even though identical twins have the same DNA, they can be surprisingly different behavior, health status, and even in physical appearance. If you think of DNA as an immense piano keyboard and our genes as keys—each key symbolizing a segment of DNA responsible for a particular note, or trait, and all the keys combining to make us who we are—the epigenetic processes determine when and how each key can be struck, changing the tune being played."

Even though each piano is constructed much the same way, an infinite number of tunes can be generated.[2]

That the environment can influence the DNA of an organism is a revolutionary idea. Many of my colleagues and I taught for a very long time that the opposite was true. Epigentics reinforces the idea that we are unique.

The second advance was that biologists were able to decipher the human genome in 2000—the genome is the whole amount of DNA in the nucleus of a human cell. We know now that DNA plays a far more important role than we thought in the second part of the twentieth century,

I. Human Diversity And Epigenetics

Recently we discovered that our choices of food or smoking could influence our DNA—and that of our future children. This is part of epigenetics, which is fast developing, although its basic idea seems heretical. After all, we have been taught that whatever choices we make, such as smoking or drinking, affect our lives, but they won't change our actual

---

2 Ibid, 64

DNA. However, this seems not to be true any more, as shown in mice[3] and in humans.[4]

In a recent human study out of 14,024 fathers, 166 said they had started smoking before age 11, just as their bodies were preparing to enter puberty.[5] When the authors of the article looked at the sons of those 166 early smokers, it turned out that their boys had significantly higher body mass indexes than other boys of the same age. That means the sons of men who smoke in pre-puberty will be at a higher risk for obesity and other health problems well into adulthood. It's very likely that these boys will also have shorter life spans. In other words, your decision to smoke at 11 years old is not only a medical mistake concerning yourself, but also a catastrophic genetic mistake concerning your children.

Even more recently, it has been observed that a father's obesity can be transmitted to his children through his sperm.[6] Epigenetic markers have been found in the sperm of obese future fathers that cannot be found in slim fathers. However, fortunately, these markers disappear in the future fathers when they lose weight.

These examples of modification are part of epigenetics. Epigenetic modifications are temporary DNA modifications due to a process called methylation.

A general example of epigenetic change is the process of cellular differentiation. A single fertilized egg—the zygote—continues to divide, the resulting daughter cells change into all the different cell types—neurons, muscle cells, skin cells,

---

3   Waterland and Jirtl. *Molecular Cell Biology*.23:5293-5300. 2003
4   Pembrey et al. *Journal Human Genetics* 14: 159-196, 2006
5   Kate Kelland Reuters.com April 7, 2014
6   Sciences et Avenir. Janvier 2016.

blood vessels, etc, by activating some genes, while inhibiting the expression of others.

II. Human Diversity and the Genomic Age

Sequencing of the human genome was accomplished as the end of the Human Genome Project (HGP). HGP was an international scientific research project funded by the U.S. government with the goal of determining the sequence of all the chemical base pairs—which make up human DNA—and of identifying and mapping all of the genes of the human genome from both a physical and functional standpoint.

At the same time, in 2000, a private company, Celera Genomics, also succeeded in deciphering the human genome. This company was formed for the purpose of generating and commercializing genomic information to accelerate the understanding of biological processes. In 1998 two of its scientists Craig Venter and Hamilton Smith were the first to sequence an entire organism's genome, that of *Haemophilus influenza*. It was logical for them to attempt to sequence the human genome that they did at a fraction of the cost of the public project. Celera Genomics' head, Craig Venter, proclaimed that his company had sequenced "the genome of three females and two males who have identified themselves as Hispanic, Asian, Caucasian, or African American" and found that "there is no way to tell one ethnicity from another."[7] This is not astonishing since ethnicity is a social construct, not a scientific one. Our genomes show evidence of extensive interbreeding, which is a very important factor in the prevention of race and ethnic formation. They have

---

7 C. Venter. Statement on decoding of genome. New *York Times*, p. D8. June 27, 2000.

been mixed and remixed with every generation so much so that the notion of any pure human population is absurd.⁸

The sequencing of the human genome holds benefits for many fields, from molecular medicine to human evolution. Genomics will soon permit physicians to treat their patients on an individual basis, rather than by the appearance of self identified ethnicity, which will be far better for the patients and physicians.

One specific, unified message accompanied the official announcement of the completion of the Human Genome Project: human beings are essentially the same. Human genetic sequence is 99.9% identical; of the 0.1% of the human genome, that varies from person to person, only 3% to 10% of that variation is associated with geographic ancestry. ⁹

With the advent of the Human Genome Project and its confirmation that human races do not exist, hopes were high that we could get rid of the concept once and for all. Contrary to this expectation though, genomic science has revived in some quarters the idea of racial categories as proxies for biological differences.¹⁰ The results of some studies have been interpreted so that they reflect and reinforce the traditional racial views of human biological diversity. Disconnects between the results and interpretation of these studies are unfortunate since they are playing an important role in the reification of race as a biological phenomenon. The study of

---

8   Craig Venter, in his footnote to Michael Yudell. *"Race Unmasked Biology and Race in Twentieth Century,"* (New York: Columbia University Press, 2014), IX and X)
9   Marcus Feldman and Richard Lewontin. Race and Ancestry, and Medicine in *Revisiting race in a genomic age.* (New Brunswick, N.J.: Rutgers University Press: 2008) 89 101.
10 People refer to races not because of any explicit commitment to a political agenda or racism, but because the hopeful curiosity that often spurs contemporary research into raced groups

human diversity would be better served by the recognition that human populations are, in reality, variably sized bio-cultural unities. Each has its own genetic idiosyncrasies, which are the results of adaptation to local circumstances and to the historical accidents of survival and proliferation.

Genomics like other biological scientific technology that had been new in the past, has led to pseudo-scientific data. This contributes to what Troy Duster, Professor of Sociology at the University of California at Berkeley and contributor to *White Washing Race: The Myth of a Color-blind Society* has identified as the molecular reinscription of race.[11] In the last few years, commerce of racialized products has appeared: there is a mania in the use of racial and ethnic categories to sell shoes, skin products, and ancestry tests. Such commercial development of genomic research related to DNA based products is a multibillion-dollar industry.[12]

There is also another financial market that accompanies this: the genetic genealogy services that contribute to the revival of the concept of race and ethnicity. People have become more and more curious about their ancestry and are willing to pay for genetic testing, which then becomes cheaper and cheaper. Unfortunately, they are not told the limits of genetic testing, which are a based on probabilistic science, not exact science. What they learn from these testing services is so general that it could apply to someone else.[13] Genetic testing like any commercial venture, sells both not only a product,

---

11 Troy Duster," Buried alive: the concept of race in science" in *Genetic Nature/Culture Anthropology and Science beyond the two-culture divide.* Alan Goodman, Deborah Heath, and Susan Lindee (Berkeley: University of California Press, 2005), 258-277.
12 Https:/en. Wikipedia org/wiki/ Human Genome Project.
13 This was the case of the author's adopted niece who wanted to know about her American Indian heritage.

but also the desire for that product. It also reinforces the concept that one belongs to a race, a nation, or continent. In the case of entitlements that are tied to race, such as affirmative action, genetic ancestral testing may inflame long-standing debates about eligibility and the social recognition of race as a class.

I understand that people are interested in their ancestry, but I doubt that genetic testing in the genomic age helps them in their endeavor. Genetic testing might indicate that your ancestors hail from Europe and/or another continent, but you may already know that from family sources.

10

The Biology of Skin Color[1]

Color is determined by just a few letters out the instruction book of life.

R.C. Cheng

Because of its visibility, skin color has been the most used biological trait to distinguish human groups and support prejudices. This is unfortunate because skin color carries no compelling, deductible conclusions regarding a person's beliefs or behavior. Only two people have been able to change their skin color, and only for a few weeks. This was done by two journalists who wanted to know how a black person feels when discriminated against. One was John Howard Griffin in 1960 and the other was Grace Halsell in 1970.

Griffin relayed his experiences in Louisiana, Mississippi, Alabama, and Georgia in his book, *Black like Me*. There, as "a temporary" black man, he was subjected to racist insults,

---

1 This chapter is an updated version of chapters on skin color that appeared in my book, The Myth of Human Races. I have included this chapter in this book because, I believe, as more and more people know about skin color, they will realize that the differences among us are striking small

squalor, violence, antagonism, and hopelessness. Grace Halsell also wrote a book about her experiences in Harlem and Mississippi as a "temporary" black woman. In her book, *Soul Sister*, she described how racism permeated every situation she encountered.

Both Griffin and Halsell changed the color of their skin, but nothing else, not the shape of their noses or their type's hair. Yet, the changing of skin color from light to dark was enough by itself to make people treat them as members of the "Negro race."

As late as 1970, racism was still so predominant that Grace Halsell had to bypass available restrooms and eating facilities to find those that were specified for her, and be called disparaging names, which referred to them as a member of a subspecies well known to be lazy, dishonest, and stupid.

One may wonder how these journalists temporarily changed their skin color from light to dark. They took psolaren, a potosentizing drug that simulates melanin production when coupled with regular and prolonged exposure to sunlight. Both became dark in a few days.[2]

To understand how it is possible, we need to know where and how melanin is formed, that is to know something about the biochemistry of melanin and the structure of the skin.

Skin color is due to a pigment melanin, which is present in very diverse groups of animals. For example, it is the black substance ejected by a squid when it is trying to get away from a predator. Melanin is a complex biological substance that is formed from the amino-acid, tyrosine. The formation takes place in a series of steps. In each step, intermediate compounds are formed. Tyrosine is first transformed into a

---

[2] If we can dark a light skin, but we do not know how to do the reverse.

diphenol, a colorless product, and then to a quinone, which is colored. An enzyme, called tyrosinase, acts as a catalyst in both of these transformations. Differences among us in skin color are due in great part, to the differences in the activity of this enzyme. In albinos, tyrosinase cannot be formed. In an individual whose skin is light, tyrosinase exists in a partially inactive form during the winter months, but throughout the rest of the year becomes active, because the sunlight gradually destroys the substances that inhibit its activity–hence the formation of a suntan. In those individuals with dark skin, tyrosinase is active throughout the entire year.

We know that melanin is formed in our skin, but where exactly and how? Thanks to electron microscopy, we now have a good idea. Human skin is structurally divided into three distinct layers. The first layer is the epidermis; the middle layer is the dermis; and the lowest layer is the subcutaneous fatty tissue. The epidermis evolves from a dense population of actively dividing cells. Its surface, a thin layer of remnant dead cells, is called the stratum corneum. Its primary function is to maintain a tough protective barrier over the entire surface of the body. The thickest part of the skin, or dermis, is composed of dense tissue containing fine blood cells, various nerves (especially those sensitive to touch), the smooth muscles that raise the hair when contracted, and a variety of specialized glands. Closely packed cells that contain considerable fat characterize the deepest layer, the subcutaneous fatty tissue. This is the layer that provides thermal insulation, reducing the loss of body heat.

Our differences in skin color are primarily due to the amount of melanin present in the epidermal layer of our skin. This pigment is synthesized in specialized cells called melanocytes, located between the dermis and the epidermis (fig.2A). These cells, which have long tentacle like projec-

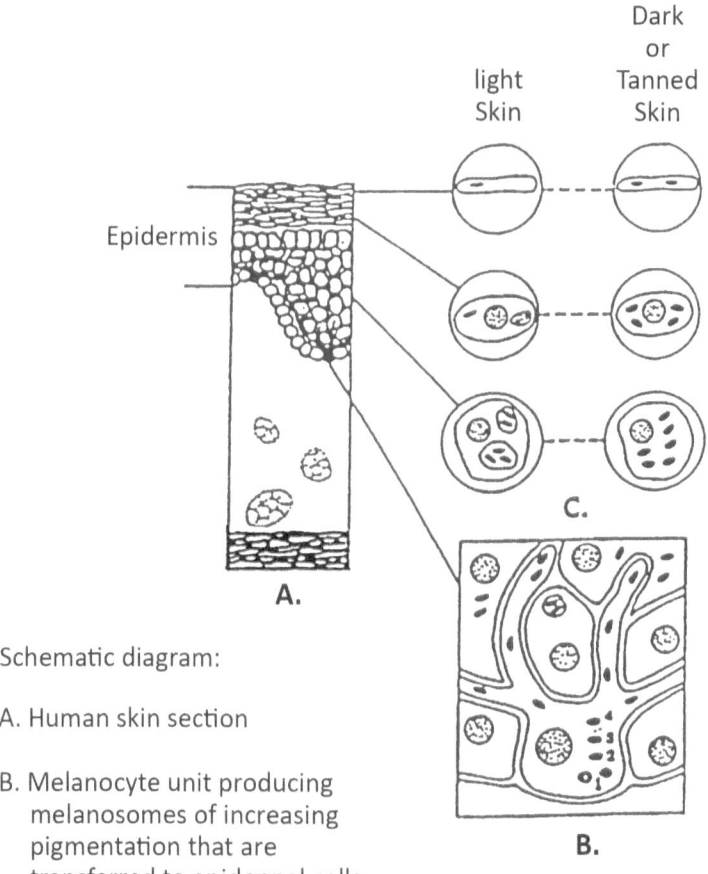

Schematic diagram:

A. Human skin section

B. Melanocyte unit producing melanosomes of increasing pigmentation that are transferred to epidennal cells.

C. Melanosomes aggregated in organelles and isolated

Figure 2

tions, inject granules of melanin into the surrounding epidermal cells (Fig.2B). There, the pigment seems to form a protective awning toward the skin's surface over each cell's nucleus.

The primary function of melanin is to shield the cell's nucleus by absorbing the ultraviolet light rays of the sun, which can change the DNA and cause mutations. Many skin cancers are the consequence of these mutations. Within the melanocytes there are even smaller structures, called melanosomes, in which melanin is produced. As the melanosomes develop, they go through various stages.

In stage one, the melanosome is spherical and has no recognizable internal structures and no melanin is present at this stage of development. In stage two, the melanosome is now oval and has an internal structure of parallel layers more or less. A small amount of melanin is evident in this stage. During stage three, the melanosome contains a larger amount of melanin. The layers that were visible in stage two are now largely hidden by melanin. Finally, in stage four, the melanosome has its maximum amount of melanin.

The melanocytes of the darker skin of Asians, American Indians, and East Indians generally contain moderately melanized stage-three and four melanosomes. On the other hand, the melanocytes of very-dark skinned people such as Africans and Australian Aborigines are usually filled with stage-four melanosomes, with only a few stage-two or three melanosomes. Before irradiation with ultraviolet light, the melanocytes of individuals with very light skin have very few melanosomes during all stages of development. Those of light-skinned individuals contain a large number of stages two and three melanosomes and may contain a few stage-four melanosomes. Once mature, the melanosomes travel to the tips of the tentacle-like projections of the melanocytes and are then engulfed by the keratinocytes, the cells that make up the stratum corneum layer-- that come in contact with them. The melanocytes then migrate with the epidermal cells towards the surface of the skin. Though there is a tendency for more deeply pigmented skin to contain larger melanosomes, the main distinctive feature that distinguishes the skin color of a dark-skinned individual from a light-skinned individual is how the melanosomes are arranged in the skin cells. Melanosomes can occur in single units if they are large enough. If they are small, about three to five tend to group together in a transparent structure like a plastic bag (fig. 2c).

Dark-skinned people generally have singular melanosomes, whereas light-skinned people, when not exposed to sunlight, generally have grouped melanosomes. However, both single and aggregated melanosomes are observed in all types of skin. Children of light-skinned and dark-skinned parents tend to have equal parts singular and grouped melanosomes. As the melanosomes travel up through the epidermis, they tend to break into amorphous melanin particles. This is particularly true of the small melanosomes that are grouped together. The disintegration of melanosomes is due to the action of very strong enzymes, and it occurs within the cells of the epidermis.

Regardless of how dark or how light our skin appears, we all have the same skin structure, we produce the same type of melanin, and we have the same number of melanocytes per unit of skin area.[3] Striking differences in skin color are caused by the amount of melanin produced by the melanocytes as well as the number and size of the melanosomes

---

3   Nina G. Jablonsliki. *Skin* (Berkeley CA. University California Press. 2006), 76.

in the surface layer of the skin. In other words, the variation in skin color does not occur at the structural level but at the functional level. It is not a question of all-or-nothing matter, but of degree.

One thing remains to be explained. The differences in skin color are extremely striking to the naked eye, yet these seemingly vast differences tend to be reduced under the microscope. To understand this paradox, let us think of a page in a newspaper with pictures. To the naked eye, the dark areas in the pictures appear fully black. However, when seen under magnification, with a simple hand lens for example, one can see the dark areas are formed by many dots with spaces in between. The larger the spots and the closer they are to each other, the blacker the image appears to the naked eye. The same is true of the skin. The dots are the melanosomes, and the larger and the closer together they are, the darker the skin appears. The fewer the melanosomes, the lighter is the skin appears, because the broader spaces between the melanosomes. Another reason naked-eye observations differ from microscopical observations is that in the first case, light is reflected from the skin, and in the second case, the light is transmitted through a thin slice of skin. The picture we get is different in each scenario. To understand this, let us take a transparent color slide and a color print of the same object. To see the object of the slide, we need to transmit light through it; however, we see the object in the color print by reflected light. The picture that we get from the slide reveals far more detail and has a far larger range of brightness than the picture from the color print. The same is true if we photograph the painted windows in a church from the inside and from the outside. We get far more details in the first case than in the second case.

The most surprising discovery that came from microscopical observations of human skin had to do with tanning, which shows, again that differences in skin color are minimal. If we look at a dark-skinned child, born in Africa and raised in Northern France, and we look at a light–skinned child, born in France, but raised in equatorial Africa, they will remain, respectively, dark and light. Therefore something fundamental must exist in their genetic make-up that determines their basic pigmentation. Scientists call this *constitutive skin color*. If our degree of pigmentation varies with our genetic constitution, it also varies with the environment. Some of us are lightly pigmented for part of the year and may become more pigmented (tanned) the rest of the year, depending upon how much our skin is exposed to the sun. Scientists call this increase in skin pigmentation above the level of natural pigmentation: *inducible skin color*.

With exposure to ultraviolet radiation, the skin of everyone, except those whose skin is very dark, becomes tanned by two distinct processes: immediate tanning is induced by the longer wavelength, UV-A; it becomes most prominent within one hour of exposure and almost completely disappears within four hours. This rapid pigmentation involves an increase of melanin in the melanosomes which transfer rapidly from the melanocytes that produced them to the surface of the skin, but there is no increase in the number of melanosomes. On the other hand, delayed tanning, which is induced by the shorter wavelength ultraviolet, called UV-B, develops four to five days after exposure to the sun's radiation. It involves new production, transfer, distribution and degradation of melanosomes. Inasmuch as new melanosomes are synthesized, an increase of tyrosinase action can be clearly demonstrated. Exposure to ultraviolet radiation also causes the epidermis to thicken, which increases tolerance to subsequent irradiation. This response results from increased cell multiplication in the upper and lower epidermis. Little is known about how this process occurs. The skin remains tanned for several weeks and offers considerable protection against further damage by sunlight. Eventually, the pigmented cells slough off, and the tan slowly fades.

I have said previously that, under the microscope, the most striking difference in people's skin pigmentation is that the melanosomes can be singular or form complexes. In dark-skinned individuals, the melanosomes mostly occur as single units surrounded by a limiting membrane. In the unexposed skin of light-skinned individuals, the melanosomes are aggregated, together as three to five particles within a limiting membrane. However, all these differences tend to disappear after exposure to ultraviolet radiation, as electron microscopic studies have shown. Tanned skin contains melanosomes in all stages of development, and many of them are found singularly, the way we find them in dark-skinned individuals (figure 4). Therefore, we should expect that the tanned skin of light-skinned individuals would be indistinguishable from the skin of dark-skinned individuals. Though this may be true under the microscope, it is not true to the naked eye. The tanned skin of a light-skinned individual never reaches the intensity of someone who is constitutively extremely dark-skinned, because it contains fewer melanosomes.[4] Nevertheless these differences are not very important, biologically.

---

4   However, the melanocytes of red-headed Europeans are small with very few dentrites (tentacles projections), a possible explanation why these people are more susceptible to sunburn.

Natural Selection and Skin Color

> Early members of the genus Homo – the ancestral stock from which all later evolved – were darkly pigmented.
>
> Nina Jablonski

It has long been known that the skin's relationship to UVR (Ultraviolet Radiation) is a complex one. This radiation both inhibits and stimulates the production of some chemical products in the skin that are very important to maintaining physiological balance in the body. Longer-wavelength ultraviolet (UVA) destroys foliate, a vitamin that is essential to the production of DNA. On the other hand, the shorter-wavelength UVB promotes the synthesis of Vitamin D in the skin, which is needed for the purpose of calcium metabolism. There is need to balance the inhibitive and stimulating effects of UVR upon the skin. Melanin is the agent playing the role. Darker skin is favored in the tropics; lighter skin in higher altitudes where the sun is not so intense. As modern human beings moved north out of Africa, they encountered strong selections against having high melanin levels in their skin. In some, slight mutations occurred in their DNA and their descendants were favored. This happened in Europe and Asia.

We believe that there has been a lightening of skin color in those living in northern latitudes because it permitted absorption of more light, which is needed for the production of vitamin D. Vitamin D controls absorption of calcium through the intestine, regulates excretion by the kidneys, and aids in the depositing of minerals in our bones. These functions are so critical that a lack of vitamin D causes a bone defect--rickets in children and osteomalacia in adults. Unlike other vitamins, it is not present enough in the normal diet. It occurs in small amounts in vegetables and in large quantities in the liver of bony fishes. Before the incorporation of the vitamin into the milk supply, the main source of Vitamin D for children was cod liver oil.[5]

If light skin is an advantage in northern latitudes, what advantage is there for dark skin in Africa? It has been discovered about twenty years ago that intense sunlight destroys another vitamin, folic acid, which is essential for proper sperm and fetal development. Hence, the selection of more melanin to protect the skin against too much sunlight must have played a role in equatorial regions.[6]

Although the correlation between skin color and latitude is not perfect, it holds well and suggests that there might have been other factors involved. Recently, it has been suggested that the skin color of the Neanderthals was also lightened. However, the mutation of the gene that produced light skin color in them was different from those in modern humans. This again suggests that a trait might result from different genetic adaptations. The darkest-skinned peoples are linked by environment of origin, not by genealogical descent. This is also true of the lightest-skinned people

Obviously, we differ in skin color, but the genetic variants affecting this trait (and facial features) are essentially

---

5   During World War II the author, who grew up in France during the German occupation, found in his home a large bottle of cod liver oil. Contrary to most of the kids, he found it not disagreeable. He did drink it from time to time and did not suffer from vitamin deficiency.
6   Nina Jablonski and George Chaplin, Skin Color, *Scientific American*. October 2002, 75-80.

meaningless in our modern world and involve only a few hundreds of the billions of nucleotides in a person's DNA. Yet, societies have built elaborate systems of privilege and control around these insignificant differences. This is well known, but what has not been emphasized is that dark-skinned women, who are victims of racism, attempt to lighten their skin tone, going as far as to put lead or arsenic based cosmetics on their skin, which causes their premature deaths through metal poisoning. On the other hand, light-skinned men and women attempt to darken their skin, because they believe that tanned skin reflects beauty and youth. Tanning became fashionable in the 1920's, when Coco-Chanel was accidentally sunburned while visiting the French Riviera. Her fans apparently liked the look and started to adopt darker skin tones themselves. Tanned skin became a trend partly because of Coco's status. In addition, Parisians fell in love with Josephine Baker, born in Louisiana, who became an icon in the night-clubs of Paris because of her "caramel-skin." Those who liked and idolized her wanted darker skin so they could be more like her. These women were two trend setters who helped begin the transformation of tanned skin being viewed as fashionable, healthy, and luxurious.

Today, research has shown that too much exposure to the sun or artificial light leads to skin cancer and premature aging. It is possible that sun tanning will one day be out of fashion, but it will not be soon because too many people expose their bodies on beaches. They think sunscreen applications prevent the damage of the sun, when in fact they might only delay them. So, the American Academy of Dermatology recommends not only the use of sunscreens, but also wearing sun- protective clothing and for many of us with light skin, avoiding the sun altogether.

From a biological point of view, it is best for us is to accept the skin color that nature has given us, and, from a social point of view, to recognize that skin color is meaningless. Skin color should never be confused with race or ethnicity. Skin color is just that; skin color.

Inheritance of Skin Color

> Genetic traits are not preformed in the sex cells, but emerge in the course of development, when potentialities determined by the genes are realized in the process of development in certain environments.
>
> Theodosius Dobzanky

Although we have known the descriptive cell biology of pigmentation for a long time, we did not know much about its inheritance, and what we thought we knew was incorrect. The genetic history of skin color is a very interesting one, because very few scientists seem to be interested in the subject.

Skin color is undoubtedly inherited, but misconceptions about it are many and have been repeated in literature. For instance, in one of the Conan Doyle's Sherlock Holmes adventures, a little girl wears a yellow mask. The reason for this strange behavior is that the mother is afraid she would lose the love of her husband if he knew that she had a child from her former husband, a black man who had died three years earlier. Showing the portrait of her deceased husband to the astonished Holmes, Watson, and her second husband, the woman said:

> This is John Hector of Atlanta and a nobler man never walked the earth. I cut myself off from my race in order to wed him, but never once while he lived for an

instant did I regret it. It was my misfortune that our only child took after his people rather than mine. It is often in such matches, and little Lucy is *darker than her father was* (italics added). But darker or fair, she is the only my own little girly and her mother's pet.[7]

There is a problem with this story. It is impossible for a child who has one very light-skinned parent (her mother), and one dark-skinned (her father) parent, to be as dark as the darker parent, let alone darker than that parent. It can be stated with certainty that no authenticated instance of such a birth has even been reported. Nor has been any well-documented case in which dark-skinned children were born to very light-skinned parents. On the contrary, in the few instances in which this was alleged to have happened, investigation revealed that the reported instance was based on hearsay and not on observed fact, or that it was the result of concealed illegitimacy involving a darkly skin pigmented person.[8]

The first studies of genetics of skin color date from 1913, Charles Davenport and Gertrude Davenport became interested in the inheritance in skin color.[9] They carried out their investigation in Jamaica and Bermuda, where marriages among dark-skinned and light skinned individuals are fairly common, since they were not forbidden as they

---

7   Arthur Conan Doyle, "The Yellow Face" in the Complete Sherlock Holmes (New York: Doubleday and Co., 1930)
8   In 1963, when the author was teaching genetics at the College of Education in Monmouth, Oregon, he received a phone call from a lawyer involved in a case of paternity. He wanted to be sure that two light-skinned people could not have a dark-skinned child.
9   Charles Davenport and Gertrude Davenport, Skin Color in Negro and White Crosses, Publication 188 (Carnegie Institute of Washington, 1913)

were in the United States. There the ancestry of the parents could be traced fairly clearly. The reader should note that, since all of us have some color, the problem that the Davenports attacked was different from the problem of inheritance of skin color in general. They were dealing with the inheritance of the differences in skin color between two human groups.

The investigators were able to find six families in which one of the grandparents was light-skinned and the other dark, and which produced thirty-two grandchildren.

They measured skin color of members by means of a color top. Colored paper disks were overlapped in such ways that varying the proportions of each color were exposed. When the top was spun, the colors seemed to blend together. By varying the proportionate amounts of color, black, yellow, red and white, the investigators could match the skin of color of persons who were studied. Data were expressed by the percentage of black on the color top. For example, five percent meant that five percent of the black disk on the color top was exposed in the matching blend.

The thirty-two offspring were classified into five different skin color categories, ranging between the extremes of the grandparents. The choice of five categories was made on the basis that five peaks of frequencies of pigmentation types seemed discernible. The investigators concluded from their data that only two pairs of genes were involved. But their analysis was not adequate for two reasons: (1) Their sample was too small, and (2) there was a considerable range of pigmentation within any given category. This meant than more than two pairs of genes had to be involved in determining the differences in skin color among these thirty-two individuals.

In 1949, Ruggles Gates[10], using the color top, studying the families of black American populations, concluded that three pairs of genes of unequal effect were involved; a similar conclusion was reached by Curt Stern in 1953.[11] To estimate the number of genes involved in skin color, Stern used an original method. He set up a series of alternative genetic models, and then compared their consequences with empirical data. Basically, he compared the observed frequency distribution for skin pigmentation in the black American population with expected frequencies based upon models using two, four, six, ten, and twenty pairs of genes. He concluded that involving three to four pairs of genes fit the observed distribution best.

But all these early studies lacked an adequate method of measuring skin color. The color top or color plate methods were unsatisfactory because they depended too heavily on the subjectivity of visual matching—the eyeball method. For example, since an exact match of the observed skin was not always found: two observers might place the same individual into different classes. A more objective and consistent evaluation of skin color can be obtained by the use of a spectrophotometer.[12] The more melanin in the skin, the less light is reflected. Spectrophotometric measurements of skin color provided evidence that there was a continuous variation in melanin amounts and suggested that many genes were in-

---

10   R.R. Gates. *Pedigrees of Negro families* (Philadelphia: Blackinston, 1949)
11   C, Stern. "Model estimates of the Frequency of White, Near White Segregants in the American Negro," Acta Genetica. stat. med 4 (1953): 281-298.
12   For a detailed history, description, and discussion of the properties of reflectance spectrophotometry, see Ashley Robin, *Biological Perspectives on human Pigmentation,* Cambridge University Press, 1991

volved in the inheritance of differences in skin color. Yet, geneticists persisted in believing that only a few pairs of genes, three to five, were sufficient to account for the differences in skin color, not only between Africans and Europeans but between Europeans and Vietnamese, Europeans and American Indians, and Europeans Australian aborigines.

Their hypothesis was widely accepted for years and publicized in textbooks about human biology. However, in 1981, Stern's method of constructing models to determine the number of genes involved in skin pigmentation was criticized by Pamela Byard and Francis Lees.[13] They pointed out that there was a basic flaw in Stern's analysis, namely, that the number of genes that one determined by his method depended on the number of classes into which the observed values were divided. For example, if we divide the range of colors into three classes, blacks, browns, and whites, the least number of pairs of genes that that can explain the data is one pair. If symbolically A1A1 determines black, A1A2 determines brown, and A2A2 determines white, then marriages between would give offspring of only three colors: black, brown, and white, in the proportion respectively 1, 2, 1. If, on the other hand, we divide the same range of pigmentation into five classes, we can explain the data by assuming two pairs of genes. This is how the Davenports explained their data. Curt Stern, who criticized their work, fell into the same trap a few years later when he arbitrarily divided the color range into eleven classes and automatically had to find that there were five pairs of genes involved. If Stern had divided the range into more classes, he would have found that

---

13   Pamela Byard and F.C. Lees. "Estimating the Number of Loci Determining Skin Color in a Hybrid Population." *Annals of Human Biology* 8 (1981): 49-58

the same data would be explained by a greater number of pairs of genes.

The relationship between the number of pairs of genes and the number of classes, in which the range of skin color of children of two medium-colored parents is divided below:

Number of pairs of genes	Number of classes
1	3
2	5
3	7
4	9
5	11
6	13
n	2n + 1

To better understand the flaw in this approach, let us turn to something with which everyone is familiar: the determination of course grades. Suppose a geneticist, who assumed that the ability to successfully pass exams is in great part genetic, wants to learn how many pairs of genes are involved in this process. He gives an exam to 5,000 students and records their scores on a percentage basis. But, teachers have to give grades and classify students into categories of A, B, C, D and F. As soon as our teacher of genetics classifies his or her students into these five categories, he or she falls into the trap of the early geneticists who researched the inheritance of skin color. Our researcher is automatically led to the conclusion that there are two and only two pairs of genes involved in the inheritance of the ability to pass exams. If the number of categories of grades increases to 13 (A+, A. A-, B+, B, B-, C+, C, C-, D+, D, D-, and F), our misguided geneticist would conclude that there are six pairs of genes involved, which is as erroneous as the two pairs he had found previously. The division into grade types is incorrect, as was the division into color types used by the Davenports and Stern. In actually, there is a continuous distribution of scores which when plotted on a graph, approaches a normal curve. This is also true for skin color data, in particular from reflectance spectrophotometry.

This semi-statistical approach to inheritance of genetics was abandoned. It is regrettable that some well-known geneticists, like Curt Stern, did not understand the nature of mathematics. It is not uncommon that famous biologists are not mathematicians. The author has learned that one of the discoverers of Mendel's paper, De Vries, did not understand mathematics, because he used the wrong mathematical method to confirm Mendel's ratios. The only reason that he comes up with them is because he knew them before hand.[14] De Vries multiplied the factors (genes) in sex cells instead of adding them up.[15] On the other hand, Mendel was a great scientist because he was not only a great botanist, but also a very good mathematician.

Knowledge about the genetics of skin pigmentation came from a very different approach. It was not until the beginning of the twenty-first century that we've had an idea about how skin pigmentation is inherited in animals, in particular the mouse. Mutations that affect mice have been studied for

---

14 Alain F, Corcos and Floyd Monaghan "Role of De Vries in the Rediscovery of Mendel's work. Was De Vries really an independent discoverer of Mendel?" *The Journal of Heredity* 76: 187-190. 1985. II Did De Vries really understand Mendel's paper?" The Journal of Heredity 78: 275-276. 1987

15 Why did De Vries do that? Possibly because plant breeders write plant crosses A x B. This is a mistake because they imply that sex cells are multiplied instead of being added.

over a century, but despite this long history, we continue to identify new genes involved in this process. They can be classified into two categories, depending on whether dermal or epidermal.[16] Among them are the KITTLG, SLC24A5, DCT, and ATRN, categories, variants of which reduce the amount of skin pigmentation and arise from a simple mutation - a change a GC pair to an AT or vice-versa. (Chapter 2).

Skin pigmentation was also studied in the golden zebra fish, a light-skinned variation of the common pet store fish. Thanks to electron microscopy, it was determined that the light color of the golden fish was due to a reduction in the number of melanosomes per cell, the average size of the melanosomes, and in the amount of pigment deposited in the average melanosomes. It was discovered that these changes were due to one specific gene, which is the same as the gene in humans. As a matter of fact, when the human gene was introduced in the DNA of the zebra fish, it changes its color from light to dark.[17]

Thanks to molecular biology, scientists were able to demonstrate that the differences in skin color between Europeans and Africans were due to at least two mutations, each in one specific gene. The mutation involves a single- base letter change in one nucleotide-a letter in the genetic code- among the three billion we have in our DNA.[18] It is hard to believe that such small changes in a few genes are responsible for the differences in skin color between Europeans and Africans. As for East Asians, they evolved light skin independently from Europeans. There is no doubt that one-letter-mutations in a

---

16 . R.K Fitch et al. (*Genes Dev.* 17: 214-228, 2003
17 Michel Balter, "Zebreafish Researchers hook for Human Skin Color." *Science* No, 5755 (16 December 2005 (1754-1755).
18 Keith C. Cheng. Demystifying Skin color and Race. "in *Racism in the 21st Century*, 3-23

few genes (at least three) are responsible for the most important cause of racism: skin color.[19] Let us take, for instance, the gene KITLG. One of its mutations reduced the skin color of those who inherited it. The variant differs from the ancestral variant by a single base-pair substitution: an A-T pair in the ancestral variant mutated into a G-C pair in the derived variant:[20]

Ancestral variant
AAAAAACTTGAA[A]GATATTATTA
TTTTTTGACCTT[T]CTATAATAAT

Derived variant
AAAAAACTTGAA[G]GATATTATTA
TTTTTTGACCTT[C]CTATAATAAT

Those who carry two copies of the ancestral variant have more pigmentation than those who carry one copy of the ancestral and one of the derived variant, and they have more pigmentation than those who carry two copies of the derived variant.

A relative lack of variation in the DNA in African populations suggests that the light- skin variants were initially rare and probably originated with a small number of people. The variants would have then rapidly increased in frequency

---

19 The sophisticated readers should read SLC24A5, a putative Cation Exchanger, Affects Pigmentation in Zebrafish and Humans. *Science 310, No 5755, (December 16,2005: 1782-1786.*

It looks like mutants of only one gene, SLC24A5, determining skin color are distributed according to ancient migration routes rather than just latitudes. Jonathan K. Pritchard. How we are evolving. *Scientific American.* October 2010., 41

20 Daniel J. Fairbanks. *Everyone is African* (Amherst: New York Prometheus Books, 2015)

as ancestors migrated out of Africa into new environments with colder temperature and higher altitudes, where light skin more readily makes Vitamin D from limited sunlight. Recently it was discovered that some Neanderthals must have been light-skinned, and red-haired like those of Irish or Scottish ancestry. Light pigmentation of the Neanderthal was due to a variant of the ancestral gene MCRI.[21]

---

21   Ibid, p.73.

11

What's the Use of Race or Ethnicity?

Despite now long –standing critiques of the concept, race continues to thrive as a category of analysis among scholars and pundits and in conventional wisdom
                           Ian Whitmarsh and David Jones

After War World II a few anthropologists argued that the biological concept of race was a myth.[1] They worked to shift focus instead onto ethnicity, a cultural category. But race persisted as a relevant category of analysis and the use of ethnicity was not alleviating the problems with race. Researchers routinely collected data (and continue to collect data) about racial and ethnic populations.[2] Both terms are used with a lot of confusion and often are linked under the pen of jour-

---

1   F.B. Livingston 1962, On the non-existence of human races. *Current Anthropology* 3: 279-281. A. Montagu. *Man's most dangerous myth. The fallacy of race* (Cleveland.: The World Publishing Company. 1962
2   Ian Whitmarsh and David S. Jones, Governance and the Uses of Race *in What's the Use of Race*. (Cambridge, MA: The MIT Press, 2010)

nalists as racial/ethnic. Meanwhile, courts use racial analyses to interpret DNA tests and predict the appearance of perpetrators, and the U.S Government (U.S. Bureau of the Census, PTO, and FDA) continues to classify its citizens by race and ethnicity according to rules that do not make biological or common sense.[3] Genetic uses of race in medicine show a similarly troubling reconfiguration of older notions of essential differences.[4] Ethnographies have shown that the strange contradictions and ambiguities also extend to the science of genetics.[5]

It is troubling because until recently, many scientists and politicians had hoped and thought that genetics would show once and for all that race has no basis in science, but this was not to be. As late as 2016, we are still asked our race (white, black, Latino, etc.) or someone classifies us himself,[6] without evidence. I suspect the classification was made on skin color and the surname of that person. In the mind of the classifier, races and ethnic groups are pure entities.

---

3   Alain F. Corcos. *The Myth of Human Races*. (East Lansing, MI: Michigan State University Press, 1997) 133-138
4   Sarah Tate and David Golsdstein. Will tomorrow's medicines work for every one? in *Revisting Race in a Genomic Age*
6, M.J. Dingel, and B.A. Koening, Tracking race in addiction research, In *Revisiting race in a genomic age* . Barbara Koenig, Sandra Soo-Jin and Sarah Richardson eds. (New Brunswick, N.J: Rutgers University Press, 2008) 172-197
5
6   Medical records have been digitalized and in mine I found that someone has classified me as "White." but wrote down that my mother language was English. I am flattered, but French is my mother language.

What's the Use of Race in Presenting Evidence in Court?

When DNA fingerprinting was introduced in the 1980s, only a few loci (gene sites) could be tested; probabilities were calculated within racial categories to increase their significance. But today genotyping technology is so sophisticated that the use of race is absolutely superfluous. Furthermore, it does not make much difference if the probability that someone has the same markers as the accuser is one in two billion African Americans versus one in three billion Europeans. Furthermore, that someone might very well live in India or Russia, not in Los Angeles, where the crime was committed. But it seems that prosecutors do not tell jury members that information. Jonathan Kahn argues that the "only thing that race adds to the presentation of such DNA evidence is race itself."[7] Another reason race should be excluded from court trials is because of the danger of infecting the proceedings with racial prejudice.

There are two major steps in using DNA for the purpose of forensic identification. First, a sample left at the crime scene by the perpetrator is compared to a sample from the suspect. Second, if there is a "match," then statistics must be used to calculate the frequency of that DNA profile in an appropriate reference population.[8] This latter step is required because it is impractical to compare the full three billion

---

7   Jonathan Kahn. What's the use of Race in presenting Forensic DNA evidence in Court? In *What's the use of Race?* Page 28

8   David Kaye. DNA evidence: Probability, population genetics and the courts. *Harvard Journal of Law and Technology*: 7 101-138 1993

nucleotide pairs (the whole genome) between two samples for forensic purposes though every person's DNA is unique. Therefore, two samples will be compared only with a limited set (usually between four and thirteen loci, or specific parts of the genome). For this practice to be effective, it is necessary to find loci that are highly variable between individuals and only test for those. In other words, there is no legal or practical justification for the continued presentation of forensic DNA evidence in terms of race. The practice can and should be ended and replaced with the use of nonracial population databases.[9]

What Is the Use of Race or Ethnicity in Medicine?

The use of racial categories in medicine assumes that they are discrete and homogenous groups. Such an assumption leads to diagnostic problems. Physicians might eliminate possible diseases or to inappropriately narrow the focus to one disease in the diagnosis of patients because of the belief that diseases only affect certain races/ethnicities.[10] A further concern regarding the analytical use of racial categories in biomedicine is the incompleteness of statistical adjustment for differences in socio-economic status among different racialized groups. In other words using racialized categories to explore disparities in health care can lead to stereotyping and prevent us from identifying the specific economic factors that

---

9   There are now private, unregulated databases culled, in part, from DNA samples people supply to genealogical websites in pursuit of their ancestry. These samples are available online to be compared with crime scene DNA without a warrant or court order
10 D.R. Williams 1997, Race and health, Basic questions, emerging directions. *Annals of Epidemiology* 7: 332-333.

contribute to group differences, such as poverty, educational underachievement, mental illness, and racism.

Years ago, medical students were taught that race was important in their diagnostics.[11] This might still be true even today, and many physicians may have never have stopped using race categories in their practice of medicine.[12] This is in spite of recent findings in genome variation science that reveal that race is often a poor proxy for genetic ancestry.[13]

Given their common, if not routine, use in the clinical encounter, it is perhaps unsurprising that racialized categories have also provided a framework for pharmacogenomics research, in which assumptions about innate differences in the efficacy and safety of drugs among different racialized groups have made them key variables in the development, prescription, and monitoring of drugs. The most prominent example of such a drug is Bidil that was approved by the U.S. Food and Drug Administration (FDA) in June 2005 for treatment of heart failure in self-identified African-Americans.[14] Though no formal claim of a race-specific effect was considered, the therapy is labeled for use in only a single human group, which was socially defined. The trial was experimentally flawed because there was no control and the experimenters seemed to have no notion that the so-called

---

11 Personal communication
12 Satel, S.L. I am a racially profiling doctor. *New York Times Magazine*, May 5, 2002
13 Duster, Buried alive. The concept of race in science. in *Genetic nature/ culture: Anthropology and science beyond the two culture*. A. Goodman, D. Heath, and M.S. Lindee eds. (Berkeley: University of California Press. 2003
14 Taylor et al. 2004. Combination of isosorbide dinitrate hydralazine in blacks with heart failure. *New England Journal of Medicine* 351_2049-2057

African American population has a large number of "white ancestors."

Although race has been thoroughly discredited as a meaningful biologic subdivision of humanity, it is still a recurring and common quantity in medical training and practice. Despite grandiose twenty-first century pronouncements of the declining significance of race and the emergence of a postracial society, it is still unremarkable to still hear clinical cases presented on the basis of race.[15] We also have a terrible time getting rid of the concept of race in medicine, in drug response, and addiction.[16]

Using race in diagnostics does little for physicians because new technologies will soon facilitate the genetic investigation between individuals, and the cost of genome analysis will be so low that every patient could afford to have one and be treated as an individual rather than as a member of a group. Race is irrelevant in medicine as it is in forensic science.

---

15   Garcia R.S. the misuse of race in medical diagnosis *Pediatrics* 113: 1394-95.
16   Molly Dingel and Barbara Koening. Tracking race in addiction Research. In *Revisiting race in a genomic age. Barbara A, Koenig, Sandra Soo-Jin Lee and Sarah D. Richardson* (New Brunswick, N.J: Rutgers University Press, 2008) 172-197

Conclusion

Is sex necessary? Not necessary, But very convenient. The world would be a much more stable, much more monotonous world without it—and we need to think only of genes to say this. The variety-generating capacity of sex is of positive value to the species that enjoys it.[1]

A.J. Muller, Nobel Prize Winner

Although science has demonstrated for the past forty years that human races do not exist, the public and the government still believe that they do. Historically, this is normal because society takes a lot of time to accept a scientific truth. After all it took centuries to realize that the earth was not the center of the universe. How many decades did it take to convince most people that smoking causes cancer?

Some scientific truths will never be accepted by everyone. There are still a lot of Americans who do not believe in biological evolution. Yet, the evolution of organisms is happening around us. For example, bacteria evolve so fast in face

---

1 As reported by Garret Hardin, Nature and Mans Fate (A Mentor Book, 1959), 186.

of antibiotics that soon we will not be able to use them to defend ourselves from bacterial diseases.[2] How long it will take to convince people that genetically modified foods are safe, that science and technology are the only way to avoid starvation, disease, and climate change?

Hence, it is highly possible that it will take a century to accept the non- existence of human races and more than a hundred years to realize that ethnicity, like human races, a social invention.[3] As to the idea that most of our ancestors did not contribute anything biological to us except life itself,[4] it might never be accepted because it is so nice to think that you are a descendant of somebody famous, forgetting that you might also be descendant of a murderer.

But the fact, most likely, is that our ancestors did not transmit their specific genes, or forms of genes (alleles), to us. That is because we have a limited number of chromosomes (23 pairs), but a theoretically infinite number of ancestors. I say theoretically because the number of ancestors increases exponentially with every generation (2^n, n being the number of generations). At one point the number of ancestors is greater than the number of people on the earth. Hence, we have common ancestors. But each of these ancestors did not transmit their specific genes to us, because of the biological phenomenon of *meiosis*. Meiosis reduces the number of

---

2   One reason is that we have been using too much antibiotics. Patients ask their physicians to give them antibiotics even when they have viral diseases. Antibiotics do not kill viruses. To keep their patients happy, physicians give them sugar pills.
3   Ethnic groups are invented and disappear. However, only one might remain: the "Blacks." In the face of racism, Jews and Muslims can change their religion, names, and customs, but Blacks cannot change their skin color.
4   After all, if any of our ancestors had not been on the earth, you would not be either.

chromosomes in the sex cells, so that after fertilization of the egg by a sperm, the number of chromosomes in the human zygote (fertilized egg) is only 23 pairs. Which chromosome we receive from each pair is distributed at random. Meiosis is the reason for our extreme diversity. We are biologically unique, because we have inherited different genes (or rather different forms of genes), and grown up in unique environments.

Some might wonder why I try to prevent people from having fun looking for their ancestry and why I attack companies that check your DNA in the search of your ancestry. Well, as a young child, I did not believe in Santa Claus. As an adult scientist, I look for the answers to strange questions, such as why you cannot find the word "heredity" in Gregor Mendel's famous paper and yet he is considered the Father of Genetics. So, I wrote a book about it with a friend.[5] I could not define race as applied to humans when I taught a course on the concept of race -- and found that no one else could either-- so I wrote a book that denies the existence of human races.[6] As far as the DNA companies, I wish they would tell their clients that it their ancestry is probable rather than certain.

Why do I believe that ethnicity is a biological myth? Race and ethnicity are often linked together and mean the same thing. In many writings, you see the expression racial/ethnic groups. If the terms mean different things, let us define them, but dictionaries are not helpful because they use these words

---

5   Alain F. Corcos and Floyd Monaghan. *Gregor Mendel Experiments on Plant Hybrids: A Guided Study.*
6   (New Brunswick, N.J: Rutgers University Press, 1993 (We also wrote a large number of articles on the subject in the Journal of Heredity and other journals) The myth of Human Races 2nd edition. (Tucson, Whearmark, 2016)

as synonyms. In any case, ethnicity and race are human inventions.

Why can't we drop the word "race" in favor of ethnic group? Many sociologists do not want us to abandon the term "race," because they are afraid that we will forget the evils of racism, but I do not believe this to be so. Racism is not confined to races. People hate others for all kinds of reasons: religious, economic, political, and sometimes for no reason at all.

Sometimes, I have to admit that it is very nice to think that some people might be related to us. In 1982, my wife and I went to Scotland—not exactly as tourists, but more like distant family members. Although she did not have any close relatives there, her maiden name is McLennan, and hence was automatically a member of a famous clan. We took the train to Inverness, Scotland, where the secretary of the McLennan clan resided, and we were treated like members of the family. We slept at the McLennan motel. Pictures were taken of the three McLennan with their tartans. One day we met the clan chief and one McLennan drove us around Loch Ness, but we did not see its monster. We had a nice vacation.

Appendix 1

Genetic and Quantitative Aspects of Genealogy

I have taught population genetics for many years. However, it is only after my retirement that I realize that this branch of genetics has its limitations. One has to do with family relationships. The standard procedure for measuring quantitavely degrees of family relationship uses Sewall Wright's method of path coefficients (Wright 1922). The measure, known as the coefficient of relationship (R) calculates the proportion of genes that two individuals have in common as a result of their genetic relationship. This works well for families 3 to 5 generations back. For example, the relation between a father and son or a mother and her daughter is ½; between a grand father and his granddaughter is ¼.. But, the method does not work well for relationships that are greater than six generations back; for example someone and his or her great-great-great-grand parent, because, as said before, it is highly possible that this ancestor did not transmit any chromosomes to this particular descendant. Wright and others dealt with genes, not chromosomes. In other words, the whole sophisticated, mathematical approach to genealogy of Sewall Wright has its limitations.

Personal example: My wife, Joanne, has a cousin Sue, in Gladwin, Michigan. However, the genetic family relationship is not close. The grandfather of Joanne and the great grandfather of Sue were brothers. R, the coefficient of relationship is only 1/96. The family relationship is excellent: Joanne and Sue have been good friends since they were young girls.

Appendix 2

What Is Mitochondrial DNA?

Although most DNA is packaged in chromosomes within the nucleus, mitochondria also have a small amount of their own DNA. This genetic material is known as mitochondrial DNA or mtDNA.

Mitochondria (illustration) are structures within cells that convert the energy from food into a form that cells can use. Each cell contains hundreds to thousands of mitochondria, which are located in the fluid (the cytoplasm) that surrounds the nucleus. Mitochondria produce energy through a process called oxidative phosphorylation. This process uses oxygen and simple sugars to create adenosine triphosphate (ATP). The cell's main energy source. A set of enzyme complexes, within the mitochondria, carry out the whole process.

In addition to energy production, mitochondria play a role in several other cellular activities. For example, mitochondria help regulate the self-destruction of cells (apoptosis). They are also necessary for the production of substances such as cholesterol and heme (a component of hemoglobin, the molecule that carries oxygen in the blood).

Mitochondrial DNA contains 37 genes, all of which are essential for normal mitochondrial function. Thirteen of these genes provide instruction for making enzymes involved in oxidative phosphorylation . The remaining genes provide instructions for making molecules called transfer RNAs and ribosome RNAs, which help assemble protein building blocks, (amino acids) into functioning proteins.

Selected Bibliography

Brodkin, Karen, *How Jews Became White Folks and What that Says about Race in America*. (New Brunswick, New Jersey: Rutgers University Press, 1998)

Corcos, Alain F. *The Myth of the Jewish Race* (Bethlehem, Lehigh University Press, 2005

Corcos, Alain F. *The Myth of Human Races* (East Lansing: Michigan State University Press, 1997, Second edition (Tucson, AZ: Wheatmark, 2016)

Davis, F. James. *Who is Black? One Nation's Definition* (University Park: The Pennsylvania State University Press, 1991)

Fairbanks, Daniel J. *Everyone* is *African: How Science explodes the Myth of Race* (New York: Prometheus Books, 2015)

Goodman, Alan H. and Thomas L. Leatherman Eds. *Building a New Biocultural Synthesis* (Ann Arbor, MI: The University Michigan Press, 2008)

Graves, Joseph L. *The Race Myth: Why We Pretend Race Exists in America* (New York: Dutton, 2004)

Graves, Joseph. *The Emperor's New Clothes: Biological Theories of*

Race at the Millennium (New Brunswick, N.J: Rutgers University Press, 2001)

Steinberg, Stephen. *The Ethnic Myth* (Boston: The Beacon Press, 1989)

Hadjor, Koff Buenor. *The Changing Face of Race: The Role of Racial Politics in Shaping Modern America* (Trenton, N.J: Africa World Press, 2007)

Hattam, Victoria, *In the Shadow of Race* (Chicago: The University of Chicago Press. 2007)

Jablonski, Nina. *Skin: a Natural History* (Berkeley, Ca: University of California Press, 2006)

Koenig, B.A, S.J. Lee, and S.S. Richardson. *Revisiting Race in a Genomic Age* (Piscataway. N.J.: Rutgers University Press, 2008)

Krimsky, Sheldon and Kathleen Sloan. *Race and the Genetic Revolution* (New York: Columbia University Press, 2011)

Lerner, Richard M. *Final Solutions. Biology, Prejudice, and Genocide* (University Park: State Universiy Press, 1992)

Lewontin, Richard, Steven Rose, and Leon Kamin. *Not in Our Genes* (New York: Pantheon Books, 1984)

Ostrer, Harry. *Legacy: A Genetic History of the Jewish People* (Oxford: Oxford University Press, 2012)

Richerson, Peter J. and Robert Boyd. *Not by Genes Alone. How Culture Transformed Human Evolution* (Chicago: The University of Chicago Press, 2006)

Ridley, Matt. *Nature via Nurture* (New York: HarperCollins Publishers, 2003)

Sand, Shlomo. *The Invention of the Jewish People* (London: Verso, 2009)

Steen, R. Grant. *DNA and Destiny: Nature and Nurture in Human Behavior* (New York" Plenum Press, 1996)

Stokes, Curtis et al. *Race in the 21st Century* (East Lansing, MI: Michigan State University Press, 2001)

Sussman, Robert Wald. *The Myth of Race: The Troubling Persistence of an Unscientific Idea* (Harvard University Press, 2014)

Tatersall, Ian and Bob DeSalle. *Race?: Debunking a Scientific Myth* (College Station, TX: Texas A.&M University Press, 2011)
Vale, Jack. R *Genes, Environment and Behavior: An Interactionist Approach (*New York: Harper and Row, Publishers, 1980)
Whitemarsh, Ian and David Jones. *What's the Use of Race?* (Cambridge: MA: The MIT Press, 2010)
Yudell, Michael. *Biology and Race in the 20th Century (*New York: Columbia University Press, 2014)

www.ingramcontent.com/pod-product-compliance
Lightning Source LLC
Chambersburg PA
CBHW030748180526
45163CB00003B/950